SOTHIS PRESS

Sothis Press aims to completely reimagine the public domain book with an emphasis on universal access to captivatingly designed print editions. We believe that technology will enable every single book ever published to be perpetually accessible in beautiful print form. Sothis is beauty in ubiquity.

www.sothis.press
1 Waterhouse Square, London, ECIN 2ST

# Big Bend

Official National Park Handbook

## National Park Service

SOTHIS PRESS

ISBN 978-1-0365-1221-7



This edition is published by
Sothis Press © 2023. The company
is registered in England and Wales.
The Company Number is 14900389.

# Big Bend

Big Bend National Park
Texas

Produced by the
Division of Publications
Harpers Ferry Center
National Park Service

U.S. Department of the Interior
Washington, D.C.

*Using This Handbook*
The major attractions at Big Bend National Park in west Texas are the Rio Grande and its canyons, the Chisos Mountains, and the Chihuahuan Desert. Part 1 of the handbook gives a brief introduction to the park and its history; Part 2 takes a close look at the area's natural history; and Part 3 presents concise travel guide and reference materials.

National Park Handbooks, compact introductions to the great natural and historic places administered by the National Park Service, are published to support the National Park Service's management programs at the parks and to promote understanding and enjoyment of the parks. Each is intended to be informative reading and a useful guide before, during, and after a park visit. More than 100 titles are in print. This is Handbook 119. You may purchase the handbooks through the mail by writing to Superintendent of Documents, U.S. Government Printing Office, Washington, DC 20402.

*Library of Congress Cataloging in Publication Data*
Main entry under title:
Big Bend.
(National park handbook; 119)
Includes index.
1. Big Bend National Park (Tex.). I. United States. National Park Service. Division of Publications. II. Title. III. Series: Handbook (United States. National Park Service. Division of Publications); 119.
F392.B53B49 917.64'932 82-600156 AACR2

1

# Welcome to Big Bend Country

# Where Rainbows Wait for Rain

Far down on the Mexican border the Rio Grande makes a great U-turn. Inside this mighty curve lies a national park and the special and spectacular section of southwest Texas known as "Big Bend Country." More than a century ago a Mexican cowboy described Big Bend as "Where the rainbows wait for the rain, and the big river is kept in a stone box, and water runs uphill and mountains float in the air, except at night when they go away to play with other mountains . . . . " This land is so vast and so wild that you can feel your human smallness and frailty. Silence takes on the quality of sound, and isolation can bring you face to face with the interdependence of all life forms.

Paradox abounds. There is killing heat and freezing cold; deadly drought and flash flood; arid lowland and moist mountain woodland; and a living river winding its way across the desert.

Spanish explorers called Big Bend the "uninhabited land," and for hundreds of years civilization passed it by on either side. Entrenched behind deep river canyons and walled in by rough and rugged mountains, this vast country remains today a world apart. Fewer than 13,000 people occupy an area about the size of Maryland, mostly in two or three towns strung along the highway to the north. Only three paved roads run south into Big Bend, and whatever route you take, you'll find yourself in country that looks less and less familiar the farther you penetrate it. Here are the landscapes, plants, and animals typical of the Chihuahuan Desert, a high dry wilderness that spills northward out of Mexico into far west Texas and southern New Mexico.

Basically, Big Bend's desert is a rolling land of creosotebush and bunch grass. But it grows gorgeous forests of giant yucca and solid stands of lechuguilla, a barbed and bladed plant found only in the Chihuahuan Desert. Big Bend's desert has living sand dunes, painted badlands, and petrified trees, and since it is a geologically young desert, its landforms stand in rugged relief. Igneous dikes march across plain and mountain like so many man-made stone walls. Chimney-tall stacks thrust up from barren flats as from a ruin. Volcanic ash heaps, white as snow, lean their concrete shoulders against maroon hills.

Since vegetation is so scant, Big Bend mountains take their shape and color from the rocks of which they are made. They loom castellated, cathedral-domed, flattopped, and razor-backed. They look red, yellow, gray, black, white, and all the shades of brown, empurpled by distance or misted over after rain in a gauzy film of green. You don't know which is more awe-inspiring, looking up or looking down, since the mountains rise with striking suddenness between the vaulted sky and the open plain. Approaching the Chisos Mountains for the first time, you can't believe that cars can breach those bastions, or that high inside there actually is a Basin where travelers have camped since people first gazed on these mountains. Undulating foothills fling themselves like breakers against the sheer rock cliffs. Standing atop the escarpment that walls up the Chisos South Rim, you see hills and mountains rolling like ocean waves far, far below, with here and there a gleam of silver where the river runs.

Big Bend's Rio Grande takes its moods from the weather, the season, the time of day, and the changing nature of its bed and banks. The river runs lavender-rose at sunset, brown between frost-reddened shrubs, shining like a tin roof under hazy skies, white as chopped ice where rapids churn, olive-green beneath the brooding cliffs of Old Mexico. Nobody knows which came first, the mountains or a through-flowing river, but for hundreds of river kilometers the Rio dodges and doubles, and where it cannot go around it rasps its way across the mountains. Deep-cut canyons alternate with narrow valleys walled in by towering cliffs. You can't get across except at a

*Preceding pages: "The uninhabited land," Spanish explorers labeled it. Its mythic topography inspired quests for lost mines and instant wealth in gold and silver. A rainbow over Cerro Castellan implies its own pot of gold.*

*Cover photo: Sunset silhouettes a century plant and lone bird through the Window, from the Basin in the Chisos Mountains.*

*Inside the gorgeous gorges of the Big Bend of the Rio Grande, the river's flow determines real time. Canoeists in fast water work the Eternal Now.*

*Other moods abound. Changing light conditions paint Santa Elena Canyon with subdued hues (opposite), then splash it with bold and saturated colors (following pages).*

handful of fords, or up steep trails at favorable stages of water. These canyons and escarpments lend Big Bend its monumental character, for as it digs, the river lays bare millions of years of Earth history. To run a desert river canyon is to penetrate the long, tortuous corridor from everlasting to everlasting: Time is here turned to stone. Imprisoned, yet wild and free, the Rio runs the ages down inside a rock-ribbed vault.

Human beings have lived in the Big Bend area for ten to twelve thousand years. The first to come were probably nomadic hunters following the big game that drifted south ahead of the last great continental ice sheet. They hunted elephant, camel, bison, pronghorn, and horses, as indicated by their kill sites discovered in the mesa and Pecos River country to the north and east of the park. But as the Earth warmed up and glaciers melted, a deadly dryness crept eastward from Mexico's Sierra Madre Occidental and engulfed Big Bend. Moisture-loving plants died out or were driven out by drought-resistant species, and as the climate and vegetation changed, so did the animals. Many Ice Age mammals perished forever and the hunters themselves seem to have disappeared.

The next people to make Big Bend their home were nomadic Indians adapted to desert life. Theirs was a follow-the-food economy, and they camped in caves and rock shelters close to such water sources as the Rio Grande and its tributaries, springs, and rock wells. They hunted desert animals for meat and skins, and ate juniper berries, prickly pears, century plant hearts, yucca blossoms, and mesquite beans. They made baskets, nets, mats, and sandals from basketgrass and the long slim leaves of the yucca. These nomadic people also disappeared, and were eventually replaced by the Jumanos, a group of farming and trading people who came to occupy the river valleys west of the park.

Cabeza de Vaca and his companions are thought to have been the first Spaniards to reach Big Bend. In 1535 they were astonished to find a farming people—probably the Jumanos at Presidio—living in houses at the junction of the two rivers. Many years were to pass before the Spanish explorers returned to scout this far country. Finding no gold or silver in this area, the Spaniards mostly ignored Big Bend.

Only a few Indian groups lived in the modern day park area, for example the Chizo, a tribe from north Central Mexico that passed its summers in the mountains north of the Rio Grande.

The Spaniards would often enslave Indians to work their mines, and the Chizos retaliated by coming down from Big Bend to raid the Spaniards. In 1644 the Chizo won a great victory, but in the end, they were driven out by a new group of Indians. These were the Mescalero Apaches, so-called because they ate the heart of the "mescal" or century plant. So fierce and skilled in battle were they that even the Spaniards fled before them. By 1720 they dominated Big Bend. Regarding themselves as mountain folk, the Apaches became the most successful desert dwellers and guerrilla fighters this country has ever known. What nature did not provide they took by raiding. Belatedly the Spaniards tried to stop the raids by building forts near major Rio Grande fords. One of these, a combination mission and presidio, was built about 1770 on the Mexican side at the park's San Vicente crossing. But the Apaches kept up the pressure, the Spaniards fled, and the fort soon lay in decay.

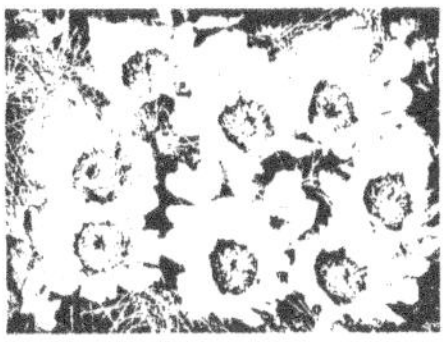

*Strawberry pitaya cactus blossoms and articulated spines pose delicate counterpoints to sweeping desert, mountain, and canyon grandeur.*

About this same time a new group of raiders, the Comanches, appeared in Big Bend. These nomadic buffalo hunters may have been the finest horsemen the world has ever seen. They ruled the south plains from Oklahoma to Texas and used Big Bend as a highroad to Mexico. For more than a hundred years, at the full of the September moon, the painted warriors crossed the Pecos and swept down past the flattopped hills and on up the long, empty, gently sloping desert floor toward the blue mountains, threading the Santiagos at Persimmon Gap where U.S. 385 now enters the park. They forded the Rio Grande at present-day Lajitas, picked Mexico clean as far south as Durango, and turned home at the end of the year. Driving captives and herds of horses and cattle northward, they probably recrossed the Rio at Paso de Chisos just to the west of Mariscal Canyon. We are told that for decades after the last Comanche left Big Bend, the great War Trail burned like a white scar across the landscape, scuffed bare by countless hooves.

Anglo-Americans took no interest in Big Bend until the Mexican War of 1848 fixed the border along

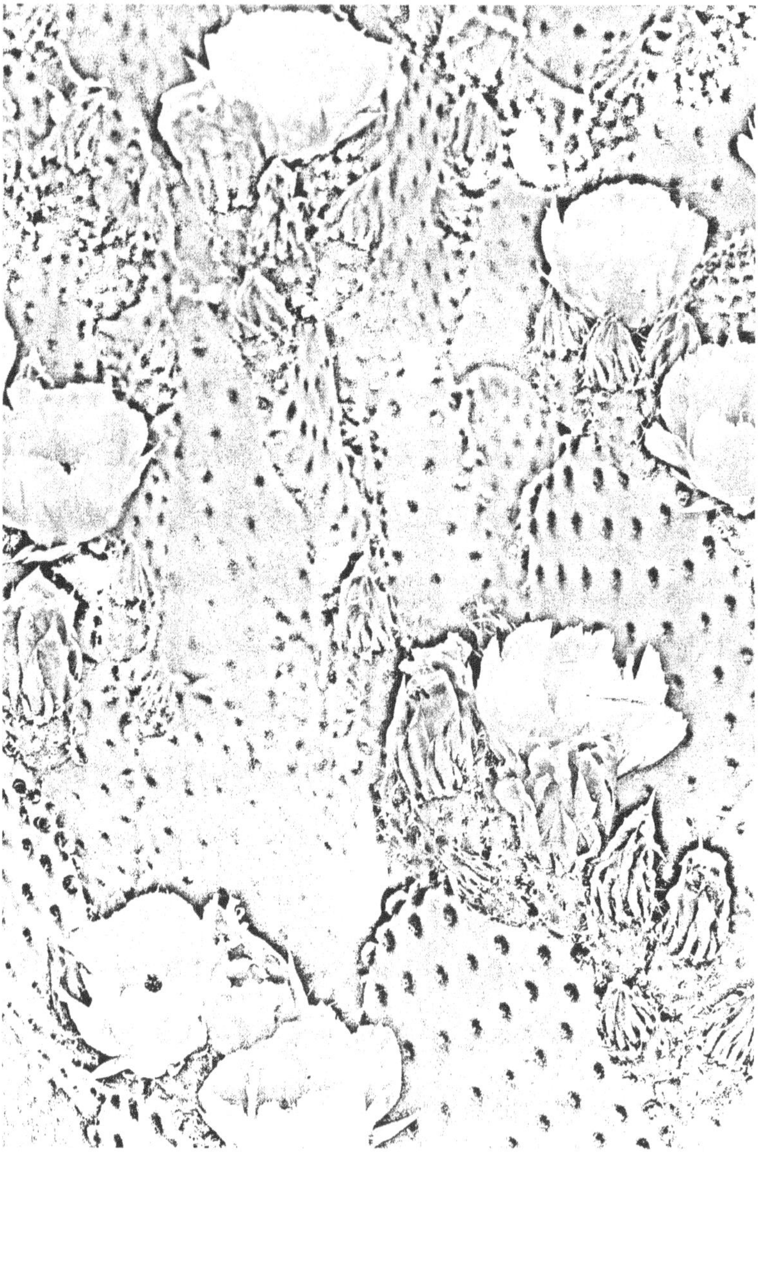

*Distinctive pads of prickly-pear cactus make this most ubiquitous of cactuses readily recognizable across its extensive range, which encompasses Southwest deserts and the Potomac River banks near Washington, D.C.*

*Following pages: The Chisos Mountains loom as an island rising above an arid Chihuahuan Desert.*

the Rio Grande. In the 1850s, two U.S. Boundary Survey teams traveled downriver by boat and mule train, and their published reports give the first scientific look at Big Bend country and its plants and wildlife. But for the next quarter-century Big Bend belonged to the Indians, and to the U.S. troopers who pushed endless patrols across its everlasting wilderness, facing sun, thirst, alkali dust, danger, and sometimes death, for $13 a month. The Mescaleros knew the country. They knew how to use its mountains, caves, canyons, and arroyos, and where to find water, wood, grass, and game. But tracked at last into their most secret and remote retreats, nothing remained for them but the reservation.

Although Big Bend Indians had long used cinnabar red in their war paint and rock paintings, it was not until the end of the last century that the commercial mining of cinnabar ore begin. To look at the ruins of Terlingua and Study Butte today, it is hard to imagine that 2,000 souls lived and worked just west of the park. Yet the Chisos Mining Company was once the world's second largest quicksilver mine, producing 100,000 flasks of mercury between 1900 and 1941. The park's own Mariscal Mine had a relatively short life and never really made money. All of the mines finally succumbed when the rich ore veins played out and the price of quicksilver fell. Similar fates overtook the copper, zinc, and lead mines that drew a couple thousand people to both sides of the river near present-day Boquillas. These mines were located in Mexico's Sierra del Carmen and the ore crossed the river to the U.S. side via a steam-driven aerial tramway. Mule-drawn wagons and trucks then hauled it over the Old Ore Road to railhead at Marathon 160 kilometers (100 miles) away.

To feed the miners, Anglos and Mexicans set up irrigated farms near Boquillas, along Castolon valley, and at Terlingua Abaja. These activities took an enormous toll on Big Bend's natural resources. Woodsmen scoured the country far and wide for timber for buildings and for firing mine furnaces. Ore train mules fed heavily on the chino grama grass. And when the mines failed and the farmers abandoned their fields, they left the land so bare that much time went by before the desert shrubs began taking over.

Ranching did not really come to Big Bend until after the Indian Wars. But once surveying parties began to locate and survey sections, cattle, sheep, goats, and horses came by trail and rail to feed upon the virgin grasslands. First-comers took up lands with permanent water; from the earliest days ranchers headquartered at Oak Spring in the Chisos. Later arrivals had to dig wells and install windmills as Sam Nail did at his ranch. The 1930s saw the end of "open range" ranching, and fencing became a prime concern for such ranchers as the Burnhams at Government Spring and Homer Wilson in the Chisos. Other ever-present problems involved water resources, drought, livestock losses from disease and predators, and remoteness from markets, schools, and doctors. Most ranchers understood the land and many loved it. They used their pastures to capacity, but they did not overstock the range until the 1940s. Then, just before the national park came into being, ruinous overgrazing all but wiped out the grasslands.

Today, Big Bend National Park sprawls across 3,205 square kilometers (1,252 square miles) inside the southernmost tip of the Bend. Even with interstate highways, park headquarters is a long way off. It is 660 kilometers (410 miles) from San Antonio to Panther Junction, 520 kilometers (323 miles) from El Paso, 173 kilometers (108 miles) from Alpine's meals and motels, 110 kilometers (68 miles) from the last community, Marathon. But the journey is well worth the effort, because the park preserves some of the nation's most dramatic land forms and rarest life forms.

The main body of the park is a great 65-kilometer (40-mile) wide trough or "sunken block" that began to subside millions of years ago, when the Mesa de Anguila and the Sierra del Carmen cracked off and slowly tilted up toward the west and east. The Rio Grande draws the park's southern boundary, slicing through three mountain ranges to form Santa Elena, Mariscal, and Boquillas Canyons. Right in the middle of the sunken block, rising higher than all the other mountains, the Chisos hang above the desert like a blue mirage.

What makes a desert, of course, is scanty precipitation. And because of the great range in altitude—from 550 meters (1,800 feet) along the river to 2,400

*Mule deer graze along the Grapevine Hills Road.*

*Following pages: A gnarled alligator juniper suggests the timeworn landscape spread below the South Rim of the Chisos.*

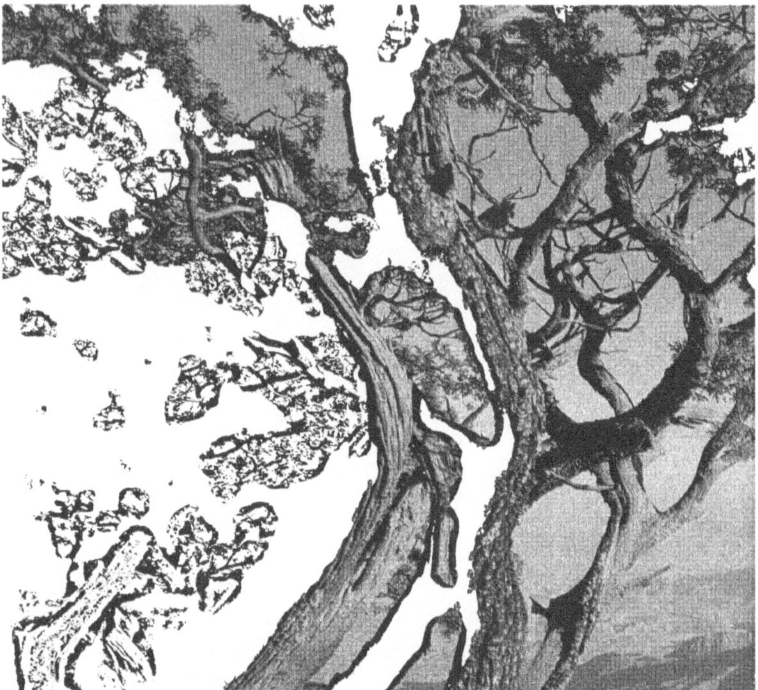

meters (7,800 feet) atop the Chisos—there is a wide variation in available moisture and in temperature throughout the park. This has produced an exceptional diversity in plant and animal habitats. Receiving less than 25 centimeters (10 inches) of rainfall in a year, almost half the park is shrub desert. This plant community begins right next to the river and runs on up to about 1,050 meters (3,500 feet). Another 49 percent of the park is desert grassland, a somewhat less dry environment that you will encounter on mesa tops and foothills to about 1,700 meters (5,500 feet). From there on up, mountain canyons and slopes may sustain typical southwestern woodlands with pinyons, junipers, and oak trees. The Chisos heights receive some 46 centimeters (18 inches) of rain per year and are considerably cooler than the desert. Consequently you will even find 325 hectares (800 acres) of forest in two or three high canyons, where towering Rocky Mountain-type trees persist from cooler, moister times. That is not all: A lush green thicket grows in a narrow belt along each bank of the Rio Grande and pushes out across the desert along creeks and arroyos. And in the river itself live creatures you wouldn't expect to find in the middle of the desert!

Big Bend National Park is home to more than 70 species of mammals, almost as many species of reptiles and amphibians, 40 or more fishes, and a fascinating host of insects and other arthropods. The wide choice of habitats makes Big Bend a birder's paradise that offers more different resident and migrant birds than any other U.S. national park. Thanks to its location, the park marks the southernmost reach of some U.S. plant and animal species, and the northernmost reach of some Mexican species. Some plants and animals found here occur nowhere else in the world.

Since its establishment in 1944, the national park has developed programs and facilities in line with its two-fold purpose of preserving and protecting natural and historic values while also enriching the lives of its visitors. Santa Elena Canyon and historic Castolon form an important sight-seeing area in the southwestern corner of the park, and the road from Panther Junction to Santa Elena is probably the most scenic in the park. The Basin lies in the heart of the Chisos Mountains. This beautiful valley has complete motel, dining, and camping facilities, and an amphitheater where park rangers give evening programs. The Rio Grande Village area on the southeast side of the park offers delightful camping and a view of Boquillas Canyon's magnificent portal. More than 175 kilometers (110 miles) of paved park roads link major sites, and well-kept hiking and nature trails take you to other spectacular areas. For properly equipped desert buffs, primitive back-country roads and trails offer exciting opportunities for true wilderness adventure.

But whether you come to the park for a weekend or a week, for vistas and views, or for a close-up look at nature and its mysteries, you will find that Big Bend is more than the sum of its parts. When the setting sun paints the Sierra del Carmen red and blue and purple, you feel both Big Bend's unity with all Earth processes and its wonderful uniqueness. In those many-colored cliffs hung above the desert, you see rainbows waiting ripe with promise for the miracle of rain.

2

# A World of Difference

*Adapting to desert dryness, creosotebushes space themselves to exploit available moisture. Their roots produce toxins that may discourage competition from other plants.*

*Preceding pages: Try to make it to the Window in the Chisos for the archetypal Big Bend sunset!*

Stand on the bald knob of Emory Peak and you'll see the Chihuahuan Desert rolled out below you with wave upon wave of mesas and mountains reaching out to the rim of the world. You can see for hundreds of kilometers in all directions. Not a house, not another human being, no living thing moves. Big Bend looks round, complete, as timeless and permanent as planet Earth itself and as beautiful and barren as the Moon. But Big Bend isn't just one world, and it isn't lifeless. It is many different worlds inhabited by countless creatures both great and small pursuing an extraordinary variety of lifestyles. These worlds may be as narrow as the mosquitofish's spring-fed pool, as wide as the cougar's hunting range, as dry as the pocket mouse's burrow, as wet as the beaver's pond, as open as the mule deer's golden grassland, or as canopied as the Colima warbler's forested canyon.

And the Big Bend world is not as changeless as it seems. Over an unthinkably long span of geologic time, and sometimes overnight, Big Bend has experienced sweeping changes that carried off whole communities of plants and animals. The great order of dinosaurs died out, and no one knows why, yet the scorpion and turtle have lived on here virtually unchanged through countless ages. Other plants and animals have staked survival on the long, slow process of adaptation to a changing environment. While one ancient lily evolved into grass, for example, another became the giant dagger we see today. Cholla cactus shades itself with thorns and the kangaroo rat manages never to take a drink.

In Big Bend as elsewhere, what animals live where is largely determined by what plants grow where. This in turn depends on such variables as the type and condition of the soil, elevation, climate, temperature, humidity, amount of cloud cover and direct sunlight, exposure to the wind, availability of water, and the drastic changes for bad and for good wrought

*Broadly speaking there are four North American deserts: the Great Basin, Mohave, Sonoran, and Chihuahuan. The park lies within the Chihuahuan. This desert is bordered on three sides by mountains; the fourth abuts vast semi-arid plains. The Sierra Madre Oriental (East Mother Range) blocks winds from the Gulf of Mexico, except as spinoffs of summer hurricanes. The Sierra Madre Occidental (West Mother Range) blocks the westerlies. How can you recognize Chihuahuan Desert? By the lechuguilla plant (see page 32), which grows only in this desert.*

by man. Yet there is nothing clear-cut or fixed about the edges of the different plant communities. The floodplain goes green or returns to dust depending on the river's rise or fall. The shrub desert, the grasslands, and the woodlands all crawl uphill or down, putting out skirmishers along their lines of march. Within the national park natural forces are once again free to shape and reshape Big Bend's different worlds. The battle seesaws back and forth between drought and ponderosa pine, tarbush and tabosa-grass, the eater and the eaten, the river and the rock, and the sun and the ageless land.

In Big Bend you can turn back your personal clock to a time when mankind was still very obviously part of nature. You can walk in the desert and drink solitude as sweet as spring water or sit on the edge of a mountain meadow knee-deep in grass. You can watch the whitetail deer drift through the forest in a silence as perfect and ethereal as song, watching you but expressing no fear. For in the park you are just one more of nature's creatures free to live and to grow in Big Bend's self-healing, life-renewing world.

For many people the spirit of the desert is embodied in the vulture tirelessly circling empty skies above a bleak and barren land, the harvester of death keeping watch over desolation. But the desert is far from lifeless or the vulture wouldn't be on patrol. The meager shrubs are miracles of adaptation and those seeming barren wastes rustle under the feet of countless busy creatures. Across the eons evolutionary selection has produced a different design for living within each species, yet all are subject to the same law.

Heat and aridity are the chief factors controlling all Chihuahuan Desert life. Most desert creatures stay in hiding during the day, keeping out of the sun in underground burrows, under rocks, or in the shrubs' sparse shade. Many birds and most larger mammals don't even visit the desert during the heat of the day. And although plants cannot crawl out from under the sun, nature has protected them by different means.

Probably the best way to see the living desert is to get out and walk and look. Study a plot of shrub desert in a single day and night. The most obvious desert dwellers, and sometimes the only living things you will see, are the plants. These vary from one stretch of desert to another because different species prefer different living conditions. But you will likely find plants in several categories, including woody and fibrous shrubs, cactuses, and other succulents. All have their own ways of resisting heat and drought, and all provide food or shelter to one or another special animal.

If success can be judged by sheer numbers, then the most successful desert shrub must be creosotebush, an evergreen bush that can make a living on the poorest and driest soils. You cannot mistake it for any other. The ground around it is apt to be bare and the individual bushes so evenly spaced that they look hand planted. This characteristic creosotebush pattern is probably caused by root competition for scant moisture. Each creosotebush has a long taproot reaching down maybe 9 meters (30 feet) to find underground water, while a network of shallow roots spreads far and wide to capture every drop of surface moisture. The plant protects itself from moisture loss by giving its dark green leaves a light-reflecting coat of resin. In the springtime, and often after rain, it bursts into brief yellow flower. It fruits in fuzzy little white balls, and you sometimes see a plant bearing both fruits and flowers.

Fortunately for the creosotebush, its taste is so unpleasant that few large animals care to eat it. But the little creosotebush grasshopper spends his whole life living and nibbling on the shrub. You'll hear him chirping away in a creosotebush, but unless he jumps you may never find him. He's a great ventriloquist who across countless generations has evolved protective coloring, the same dark green as creosote leaves, marked with the same red and white of its little stems and fruits. You may never see the mottled gray and black walking-stick insect either, who sticks his front legs straight out in front of him to look exactly like a woody creosotebush twig. Creosotebush holds the desert soil as blowing sands heap hummocks around its stems, and these make favorite burrowing sites for all sorts of little desert rodents and reptiles. Look under almost any creosotebush, and you will see their holes. You may even see a busy line of ants taking bits of creosote leaf and fruit to an underground nest.

Another curious woody shrub resembles a sheaf of coachwhips. If you see ocotillo after rain, it will look

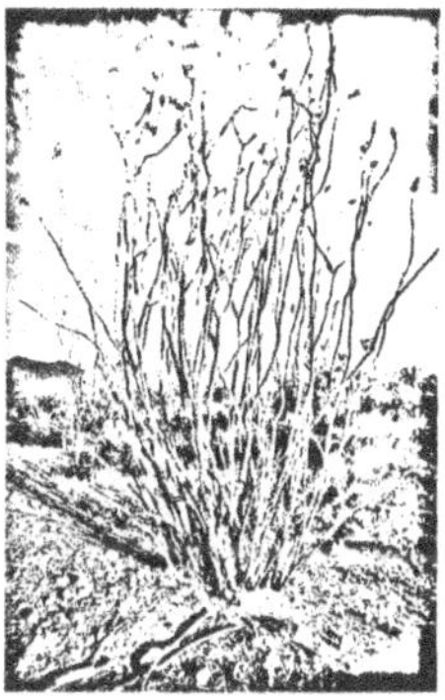

*The ocotillo also goes by the name coachwhip because it so often looks like a bunch of buggy whips stuck in the ground. However, in the springtime following a wet winter, those dead-looking stalks are adorned with green leaves and topped by brilliant red flower clusters. The ocotillo is common throughout both the Chihuahuan and Sonoran Deserts.*

like a green fountain. If you see it during a dry spell, you may think it is dead. Not so. Ocotillo puts on a fresh close-fitting suit of green leaves whenever it rains. Then as soil and air dry out, it sheds its leaves right down to the bare brown stems. This cuts back on the plant's water needs. Moisture loss is further reduced by resinous cells that form the inner bark. In springtime the tip of each ocotillo wand burns with a cluster of scarlet flowers.

Cactuses have done away with leaves altogether, thus reducing their surface area and cutting down on moisture loss. Although they look dry and forbidding, inside that harsh exterior their flesh is moist and succulent. The food-making function has been taken over by the thick, green, wax-covered stems. And since the stems are also used to store water, they have ballooned into a weird and wonderful assortment of shapes and sizes. You'll find the mound-building strawberry cactus with its mass of finger-like heads, the Texas rainbow cactus with its small group of cylindrical heads topped by bright yellow flowers, the tuber-like living rock, the dog and cane cholla, and the great sprawling pricklypear that lifts its beavertails from desert flat to mountaintop.

Cactuses come heavily armed with spines so cleverly shaped that they are called "fish-hook," "eagle's claw," and "horse crippler." These spines serve a double purpose: By building a lattice work around the stem of a cactus they shade it from the sun, and in many cases they make the cactus too prickly for animals to eat. Some animals have learned to use cactus spines for their own protection. The big, ratchet-voiced cactus wren likes to build its nest in the densely spined cholla, and the packrat often piles pricklypear pads in its nest area.

Pricklypear is the commonest cactus in the park and also the easiest to identify. Purple-tinged pricklypear is just what its name suggests, and so is the brown-spine pricklypear. Blind pricklypear looks as if it has no thorns, but if you touch one of the velvety buttons on a pad, you will pick up a fingerful of almost invisible, but highly irritating little spines. Engelmann pricklypear is the most abundant species. And to human taste its fruits are delicious, although the tiny glochids, barbed spines, can hurt the mouth. Many desert creatures eat pricklypear: Flies, bees, and butterflies come to feast at the

## Cactuses

*Fish-hook cactus*

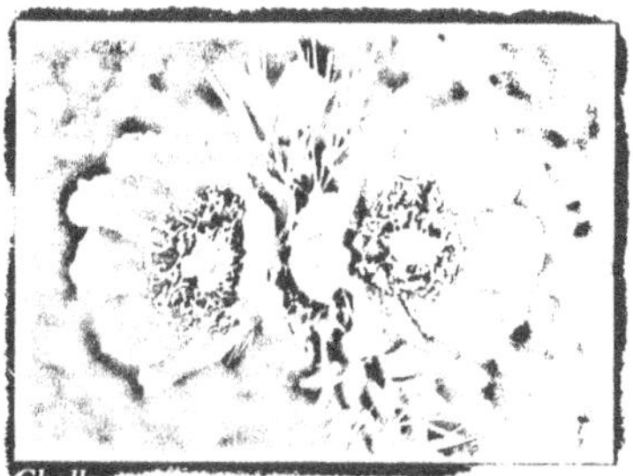

*Cholla*

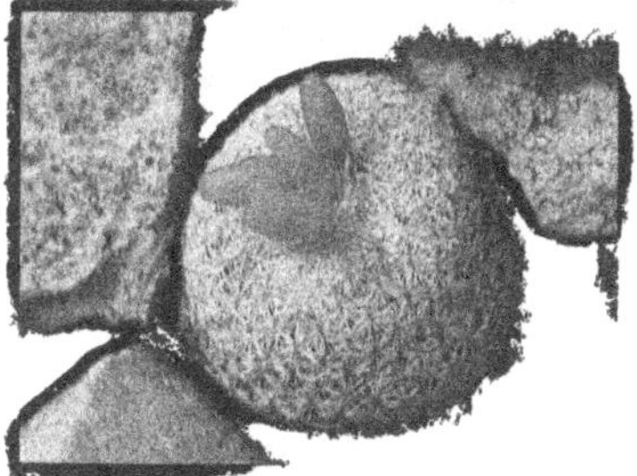

*Button cactus*

*Pricklypear*

*Strawberry pitaya*

*Eagle's claw*

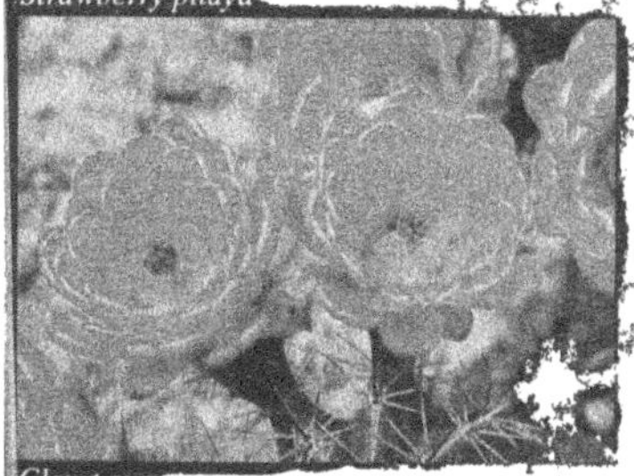

*Claret cup*

*Rainbow cactus*

showy blossoms; birds, coyotes, peccaries, and deer eat the reddish fruits; small rodents reach between the spines to nibble on the juicy pads. In times of drought ranchers burn off the spines and feed the pads to cattle. And people lost in the desert can do as the Indians did—peel, cut, or roast the skin off the pads and the flesh yields both food and moisture. Some pricklypears are too bitter to eat, however.

One of the most interesting plants in the Chihuahuan Desert is a sturdy bunch of blades called lechuguilla. This fiercely spined agave lives nowhere else in the world. When Cabeza de Vaca crossed Big Bend in 1535, the lechuguilla grew so thick that he didn't dare walk at night. Today you find it growing singly or in colonies from the shrub desert clear up into the Chisos woodlands, its needle points still menacing hikers, horses, and deer.

Lechuguilla is a fiber plant that keeps its juicy parts underground until it blooms, which it does only once a lifetime, after ten to fifteen years. The bloom stalk shoots up like a giant asparagus spear maybe four meters (15 feet) tall, flowering from the bottom up in close-packed purplish or yellowish blooms. Then the whole plant perishes by degrees. You may find a lechuguilla whose blades have died and dried while the bloomstalk is still moist and green. Eventually the bloomstalk, too, will turn into wood strong enough for a deer to lean against when rubbing velvet from his antlers.

Lechuguilla reproduces both by seeds and by rhizomes, and you sometimes find tiny new rosettes breaking ground on the runners of a mature plant. Peccaries often root up the juicy lechuguilla rhizomes, while mule deer relish the tender bloomstalk, munching it much as a cow chews a stalk of corn. Pocket gophers eat the core right out of a standing plant by tunneling underground.

The kinds of animals you meet in the desert will differ with the time of day and the time of year. They must find food and moisture, mate, and raise their young without exposing themselves to killing heat and the risk of dehydration. Insects, spiders, scorpions, and reptiles all derive their body temperature from their surroundings. This is why they stiffen up to the point of helplessness when it's cold, and why crawling on a super-hot surface will kill them in short order. Their temperature regulation problems

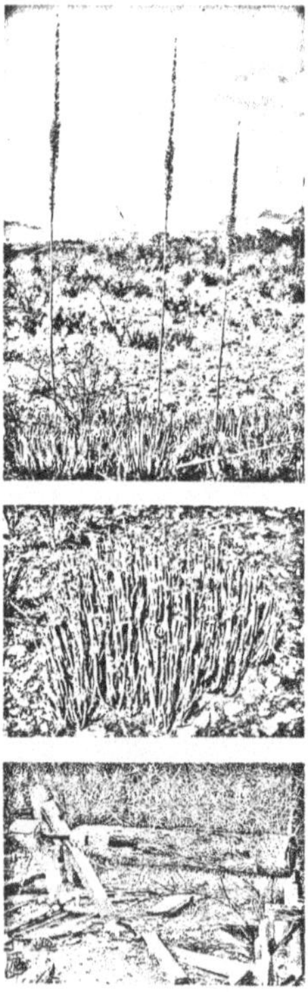

*Coarse strong fibers of the lechuguilla plant (top) were extracted by machine (bottom) for use in matting, ropes, bags, and household items. The candelilla or wax plant (middle) has been used in manufacturing waxes, polishes, chewing gum, phonograph records, and candles. In the rainy season the stem fills with milky sap. In the dry season this sap coats the stem as wax by evaporation. The wax protects the plant from drought.*

are compounded by conditions in the desert, and most of them cope by modifying their behavior.

In the early morning you may see grasshoppers sunning themselves on a rock as rattlesnakes will do. They line up broadside to the sun's rays, raising their wings and lowering their legs to expose their abdomens directly to the sun's warmth. By noonday they line up parallel to the sun's rays to minimize heat absorption, and they will seek shade. Also, instead of hopping over the ground, many Big Bend grasshoppers live in and fly from bush to bush. The surface of the desert may be 20 to 25 degrees Celsius (40 to 50°F) hotter than is the air just over a meter (4 feet) above the ground. This is why in the daytime you will mostly see only flying insects, such as butterflies, grasshoppers, true bugs, true flies, and bees, and why so many crawling insects stay hidden during the heat of the day. However, there are some curious exceptions: The darkling beetle scurries about over the sand throughout the day. An air-filled space under its hard outer wing-covers acts as a kind of insulation between the back and abdomen. Some darkling beetles also raise the abdomen at an angle of about 45 degrees. Speed is of the essence for this little scavenger as it scurries from cover to shade.

Some grasshoppers are ground dwellers that have lived on the desert pavement so long that they even look like stones. Many come in conventional grasshopper shape but are mottled in shades of gray and mauve. The toadhopper that inhabits wash bottoms and rocky areas has taken on the color, shape, and texture of rock. Fat and squatty, he will camouflage himself, tucking his antenna right down in front of his face and pulling his legs in close to his body. You can't even see him when you know he is there. By comparison, the lubber grasshopper advertises his presence. He is a large black beast gaudily marked in coral-snake red and yellow. Apparently these colors warn predators that the lubber is distasteful. He is out and about from late morning on.

Also seen in broad daylight is the worm-like millipede rippling its way across the desert pavement. This maroon-colored plant eater has up to 200 legs arranged in short double pairs along a 13-centimeter (5-inch), many-segmented body. He isn't poisonous and won't sting or bite, but he may emit a substance lethal enough to kill other insects in a confined area.

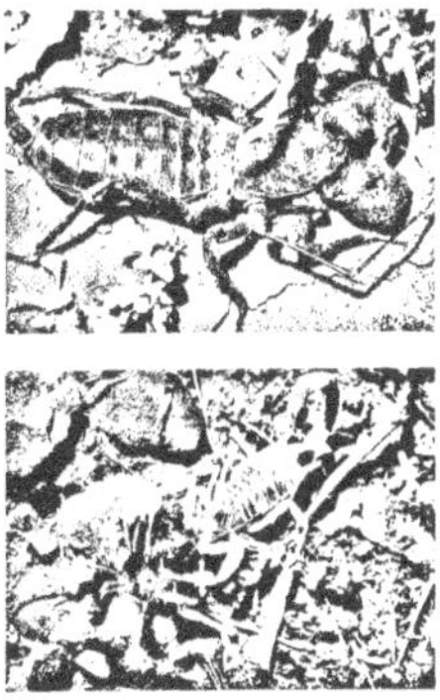

*Eleven species of stinging scorpions (bottom) live in the park. Coloration varies from dull-cream through brown to shiny black. Whiptail scorpions (top), which are not true scorpions, have no stinger. They pursue insects and other invertebrates and kill them with powerful pincers.*

Of course, the chief daily events of life in the desert are eating and being eaten, and predators that favor a certain diet make it their business to be out when their kind of dinner is around. Thus the grasshopper-eating lizards brave the daytime heat to do their hunting. Most often seen is the quick moving western whiptail. The greenish collared lizard may be seen racing along on his hind legs like a miniature dinosaur. Lizards are about the biggest ground dwellers you scare up on a noonday walk—unless you happen on a lizard-eater like the big, pink western coachwhip snake.

At twilight you become most aware of the desert's residents. The coolness brings them out. Some must hurry and eat before it gets full dark, while others have the whole night ahead of them. At first you may sense the desert's coming-to-life more by listening than by looking. You hear the lesser nighthawk trilling like a toad. Then, without warning, the whole desert begins to sing, as katydids, grasshoppers, and crickets join in a tapestry of sound so rich you can almost touch it.

Soon the desert cottontail creeps from his thicket to nibble pricklypear fruit. He stays close to home and prefers brushy terrain. The blacktailed jackrabbit passes the day in a form, a basin scratched out beneath some bush. He can cover the ground in enormous jumps, and his megaphone ears help cool him by dissipating body heat. The desert mule deer, another blacktailed, long-eared browser, may also appear at dusk to forage mesquite and lechuguilla. And a band of peccaries—or javelinas as they are also called—may rattle through the brush. They have a great fondness for pricklypear and their mouths are so tough they eat roots, fruits, pads, spines, and all. The ferocity of these wild pig relatives is more fiction than fact. If you meet one face to face, he may take a few steps toward you, but not out of meanness. He's nearsighted!

Evening can linger a long time in the desert and night can strike quickly as a cat's paw. You watch the sun go down, turning the clouds above the Chisos red, painting Sierra del Carmen crimson, while a single golden shaft breaks through the clouds and hits El Pico like a spotlight. Then the sky goes smoky blue and mauve over the eastern mountains, and the clouds to the west turn ashen as burned out coals.

## Lizards

Collared lizard (male)

Collared lizard (pregnant female)

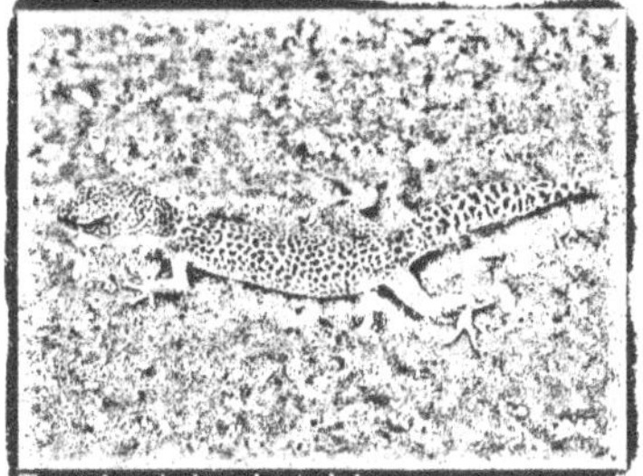

Texas banded gecko (adult)

Texas banded gecko (young)

Texas alligator lizard

Marbled whiptail

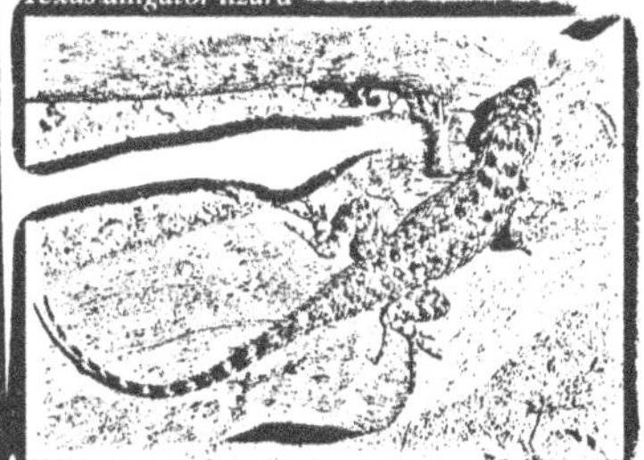

Twin-spotted spiny lizard

Big Bend gecko

# Snakes

*Western hognose snake*

*Western coachwhip*

*Mexican milk snake*

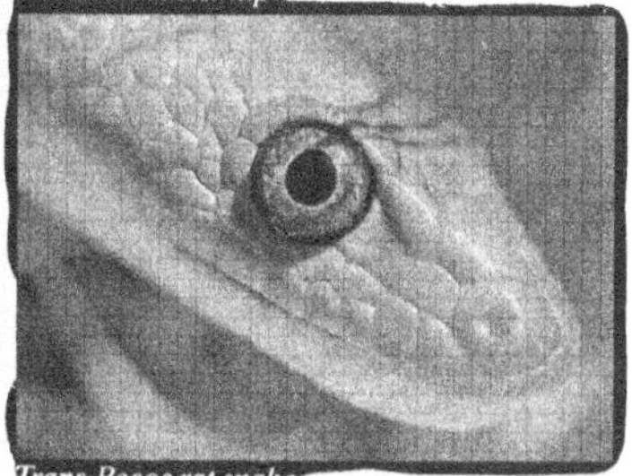

*Trans-Pecos rat snake*

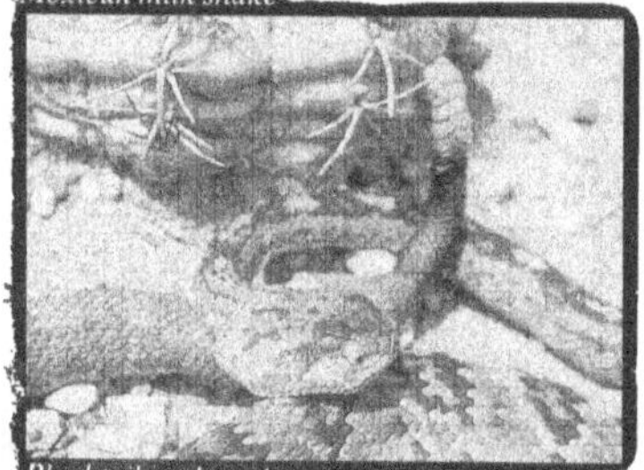

*Blacktail rattlesnake*

*Western diamondback rattlesnake*

*Trans-Pecos blind snake*

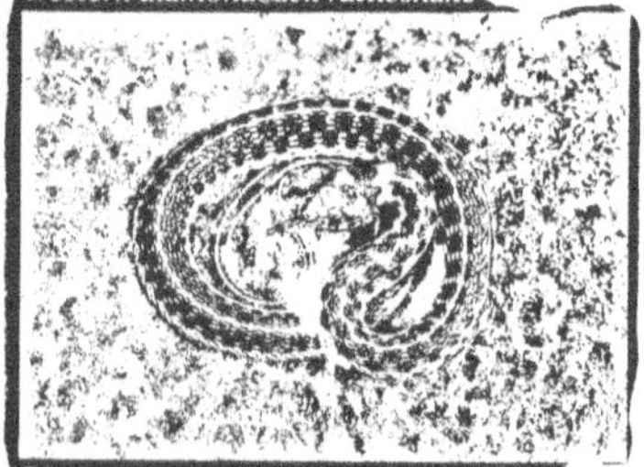

*Blackneck garter snake*

*Baird's rat snake*

*Mohave rattlesnake*

*Big Bend patchnose snake*

*Longnose snake*

*Texas lyre snake*

*Black-hooded snake*

*Glossy snake*

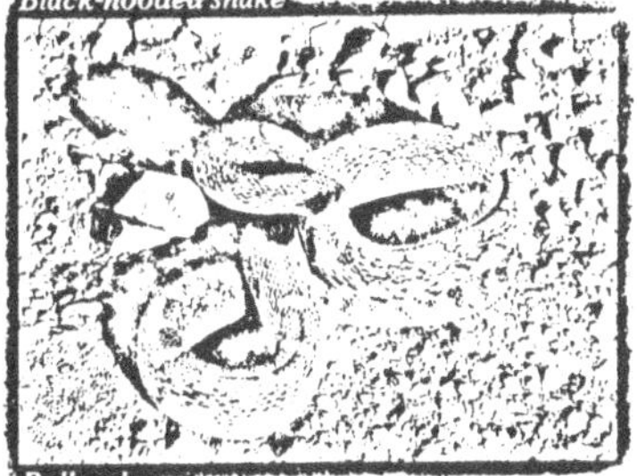

*Bullsnake*

## Two Non-Drinkers

The kangaroo rat and road-runner exemplify adaptations for desert living. Neither drinks water, as a rule. The roadrunner gets its moisture largely from its omnivorous diet, which includes lizards and small rattlesnakes. It kills them with stunning blows of its beak. Its characteristic X-track provides good traction in sand. Agile and nimble, this 60-centimeter- (2-foot) long bird can fly, but it prefers to run, at up to 32 kph (20 mph). Mexicans call the roadrunner *paisano*, "fellow countryman."

The kangaroo rat metabolizes both energy and moisture from seeds that contain less

than 4 percent water. It has no sweat glands and cools itself by breathing. Its nasal passages, cooler than the rest of its body, condense breath moisture for retention. Its kidneys, among the most efficient in the animal world, excrete uric wastes as a concentrated paste, not as liquid, saving further precious water. Its deep burrow has a year-round relative humidity between 30 and 50 percent.

These rodents sometimes fight with each other, leaping high into the air and striking at each other with their strong hind legs.

You can see the nighthawk now against the pale and pearly afterlight, and a star pops out, then another and another. Suddenly, more stars seem to be twinkling than can possibly exist in the universe. In the absence of man-made light they are an overwhelming presence. The Milky Way stretches from horizon to horizon. Who can believe that our Sun is just a middle-sized star, and planet Earth a mere speck spinning on the fringes of that gorgeous luminosity?

On a night of no moon, in the mist of starlight, the mule deer may stay active until dawn. On mild, windless nights the hunters and the hunted come out in full force: insect-eating scorpions, tarantulas, and wolf spiders; seed-eating pocket mice and kangaroo rats; rodent-eating snakes, badgers, and owls. What a hurrying and scurrying, what popping up from holes and burrows, what slithering and digging, what squeaks and shrieks, what patient waiting in ambush. And by what ingenious means do the hunters find the hunted in the dark! Beep-beeping bats locate insects and avoid obstacles by bouncing sound waves, imperceptible to humans, off objects as they fly. The female katydid wears her ears on her knees; by waving her front legs she zeroes in on the male's mating call. Cold-blooded rattlers heat-sense warm-blooded rats and mice. And just as an astronomer opens the aperture on his telescope, so the owl at night widens his enormous eyes for light from far off stars.

Toward dawn the morning star burns like a lamp in the east, and gradually, a pale flush spreads upward from the crest of the Sierra del Carmen. A bank of clouds hangs off the Fronteriza, and as the overhead stars wink out and the morning star burns on, the pale glow turns peach and seeps higher. Just enough air stirs to shake the mesquite. A waking bird emits one cluck. Soon the clouds below the Fronteriza go salmon pink and flare with internal fire. As the sun tops the Sierra del Carmen and spills a glare sharp as ice shards over the desert, you hear a distant bark. One yap, two, a soprano howl, an alto tremulo, then chord upon chord in wild and worshipful sounding chorus. Somewhere in the ruddy hills a pack of coyotes seems to sing the sun up.

*Desert flowering plants adorn a mudflat almost as metaphors of patience. The secret lies with seeds that have adapted to remain dormant for years, if necessary, until enough rain falls to bypass their germination inhibitors.*

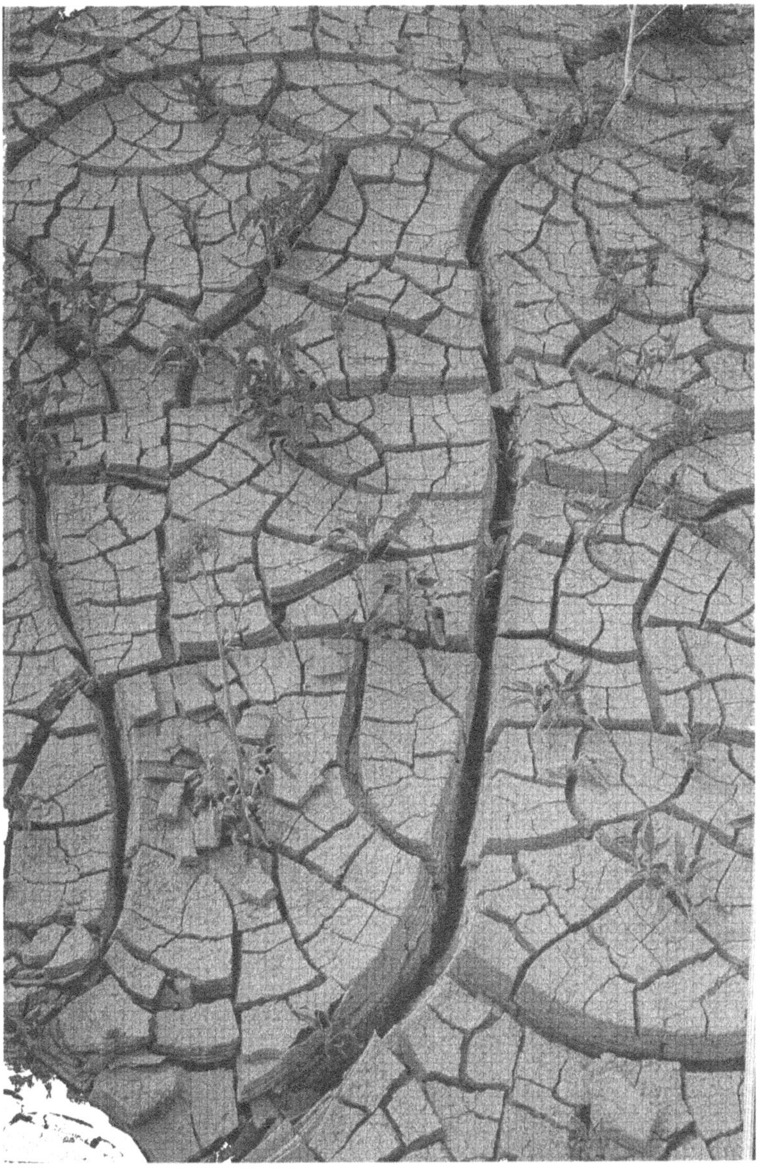

# Waterholes, Springs, and the Fifth Season

*When rains raise the water level, animals drink from natural tanks such as Ernst Tinaja. Tinajas can also be death traps when the water level falls so low that animals can't climb back out. Mountain lion claws have etched desperation into the rims of some.*

One of the most astonishing sounds in the desert is that of trickling water. One of the happiest desert sights is a pool dancing with aquatic creatures. Who can believe it: Tadpoles darting about, water striders dimpling the surface, blue darners stitching zig-zags through the air and dipping the tips of their abdomens into the water? What a celebration of life in the midst of apparent lifelessness.

Water is the single most important need of almost all life forms in the desert. The larger mammals, many birds, and some insects must drink daily to survive. Some amphibians and arthropods must spend at least part of their lives in the water. Each waterhole is a little oasis supporting its community of plants and animals, and drawing from the outside world a thirsty parade of creatures that comes to it to sustain life.

Apart from the river, there are at least 180 springs, seeps, and wells in the park that serve as wildlife watering places. Most of these are springs located within the grasslands on the lower slopes of the Chisos Mountains. Springs differ greatly, ranging from a seep with 0.5 centimeters (0.25 inches) of water standing in the grass, to a 25-centimeter (10-inch) deep pool the size of a table top, to a string of pools connected by a flowing stream. Since springs depend for their flow on water seeping through the ground, and since this in turn depends on rainfall, the amount of water found at a spring may vary greatly from season to season and from year to year. Other crucial factors are evaporation and the water consumption and retention properties of the spring's plant life.

You can see most springs a long way off. They stand out like timbered islands in an ocean, with tall cottonwoods, willows, and honey mesquite, and man-high thickets of thorny acacia festooned in silver showers of virgin's bower. Dozens of little rodent holes perforate the ground among the roots and the

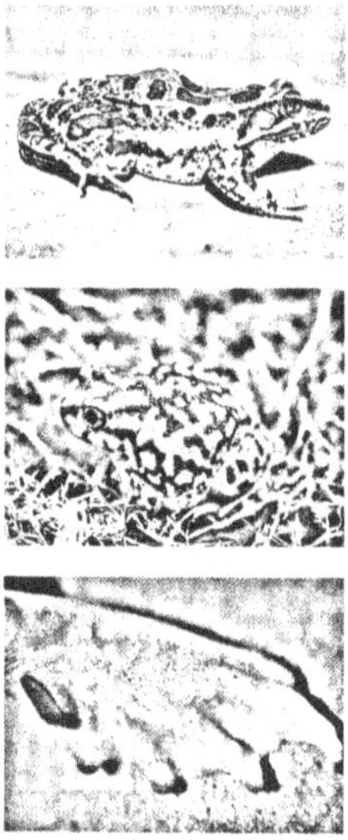

*Desert amphibians? Leopard frogs (top) live along the river and near ponds and springs. Couch's spadefoot toad (middle) evades drought by burrowing with specially adapted hind feet (bottom). When rains come, the toads move to the nearest puddle and mate. Their eggs hatch six times faster than those of garden toads and the tadpoles quadruple their birth weight by the second evening of life. With luck some mature before the puddle evaporates—and dig in to await another wet spell.*

tall grasses quiver with furtive comings and goings. Life at such a spring follows a regular pattern from dawn to dusk, although it may actually be busiest at night when most desert creatures are abroad.

At first daylight four or five redheaded turkey vultures stir in the cottonwoods where they have spent the night. They shrug their black shoulders and wait for the sun and the thermals to rise. An early blacktailed gnatcatcher chases a late moth, but the moth proves the better acrobat and makes it to safety in the thicket. Doves leave the ground with a flutter of white-barred wings and level off across the desert. By following the game trails to water, you can read the sign of nighttime visitors: The cloven-hoofed track of peccaries imprinted in the ooze, cigar-shaped coyote scat complete with fur, the flat-footed print of a striped skunk, and the larger cloven hoofprints of mule deer.

Soon it is full morning with flies biting, lizards scuttling, and butterflies feeding in jackass clover. By noonday a brisk breeze is shaking the cottonwood leaves, producing a sound like rushing water, and two ravens have come to croak in a little mesquite. Now they fly, with the sun striking silver from jet feathers. They circle the oasis, flapping and soaring, driving their shadows below them over the ground.

Here on a willow trunk is a life-and-death contest. Rubbed raw by the branch of a neighboring tree, the willow is exuding sap from a saucer-sized wound. Drawn to the sap, six butterflies stand on the damp spot peacefully feeding, slowly opening and closing their wings. All at once a mantidfly pounces from ambush and grabs at a butterfly with his clawed front legs. The butterfly leaps like a scared horse, and in reaction the whole group takes to the air. But in a moment they settle back down, roll out their tongues like party toys, and begin to sip. Another fierce lunge by the mantidfly, another scattering of butterflies. And all the time you can hear the tick-tick-tick of a beetle boring a burrow in the diseased wood.

As evening comes on, the doves come in from the desert, flying low along the line of seepage. The vultures return to roost, lazily circling the cottonwood's crown. While it is still light the butterflies seek cover in the cottonwood leaves. As it gets dark the moths come out, and after them the bats, beep-beeping as they cut erratic patterns through the dusky air. Soon the breeze will die down, and the starlit night will throb with the long drawn trill of tree crickets. In the wee small hours there will be no sound, no breath of air or outward sign of life. Then suddenly along a sandy trail moves a blackness shaped like a high-backed child's chair. It is a striped skunk, tail-high, come to take its turn at the waterhole.

Few and far between are the springs with sufficient flow to send a brook singing down a ravine. But such a place is Glenn Spring, the chief spring along a dry draw that starts in the Chisos and cuts deeply through many-colored clays as it crosses the desert. Historic photos show Glenn Spring enclosed within a man-made rock wall. Today you cannot even find the source, so thick is the tangle of tules and cane grown up around it. The flow from Glenn Spring trickles down the draw about 1.5 kilometers (1 mile), collecting in pools and gurgling over rocks before it goes underground. Some of the pools are crystal clear, and some are black with the acids of plant decay. Deeper pools are fern-green with algae. Little black snails harvest algae on the rocks and leopard frogs croak and plop, so quick to hide among the reeds that you can hardly find them. These slim, spotted amphibians, insect feeders, mate in water. Their larval young, free-swimming tadpoles, must live in water, feeding on microscopic organisms until they grow lungs and legs for life ashore.

The tadpole itself falls prey to giant water bugs, air-breathing water dwellers that are also strong fliers. The water boatman is a vegetarian who sculls about from one underwater plant to another. You can hardly tell him from the backswimmer except that the latter swings his oars upside down and spends much time on the surface hanging head down, the better to spy the aquatic insects upon which he feeds. The water strider is another hunter, but this spider-legged semi-aquatic skates atop the water, seeking terrestrial insects that have dropped onto the surface. Just as the birds and bats eat different foods at different feeding levels, so do the creatures that inhabit a pool, be it only centimeters deep.

And the creatures above the pool: The damselfly alights on a reed and rests with its transparent, net-like wings closed above its slim body. The stouter-bodied dragonfly rests with its wings outstretched and likes to fly in tandem. Both of these aerial

*The javelina, or peccary, smells like a skunk. This nighttime wanderer uses the scent for territorial marking, not defense. Curious and shortsighted, javelinas might approach a hiker—not to attack, but to investigate.*

*This flash flood (above) washed out a portion of the Maverick Road. Flash floods can be killers to the unwary. They can sweep down on you from storms you never saw or heard.*

*Cottonwood Creek's wide bed (opposite) suggests that it, too, knows rage. Low water levels favor algae growths whose colors mirror the cottonwoods' refreshing verdure overhead.*

beauties must lay their eggs in water, and their larvae are fully aquatic predators that breathe with gills like fish.

Many of the same water insects inhabit yet another type of waterhole, the tinaja, a natural pothole that traps rain or runoff in solid rock. Dependent on rainfall, tinajas often dry up, yet they may be the only water source over a large area. If a tinaja is deep enough it may survive evaporation, but the water may shrink back so far below the lip of the bowl that animals cannot reach it. A cougar once drowned in a tinaja here because it could not climb out again. Tinajas may also turn into death traps for the plants and insects that inhabit them. In a well-balanced pool the algae create the oxygen and food that aquatic creatures need, but as the pool dries up there is less and less oxygen and the products of decay become concentrated. At last these become so poisonous that the reproductive engine cuts off and the pool is literally dead. But even a dead pool may be a source of life to outside animals.

Ernst Tinaja is a good site for watching desert wildlife. It lies in a rocky, canyon-like drainage near the Old Ore Road. Though the upper tank measures 6 by 9 meters (20 by 30 feet) animals may not be able to use it because when the water is 3 meters (10 feet) deep it lies more than one-half meter (2 feet) below the edge. But mule deer and javelina frequent the smaller pools, which likely hold algae and a roster of aquatic life.

Other important tinajas may be found on Mesa de Anguila. The mesa top has a maze of trails leading to and from tinajas that have served as a focus of life across countless centuries. You can find Indian shelters in the form of overhanging cliffs up and down a canyon, with a permanent tinaja right in the middle.

Like the so-called lower animals mankind has long been dependent on waterholes. Since the first prehistoric Indians came to Big Bend, people have lived beside springs and tinajas. And what a pleasant prospect you still find from the sooted rock shelters above Croton Springs as you look out across the grasslands and the tules at the spring, toward the crenelated wall of the Chisos. Rounded red boulders beside the spring contain age-old mortar holes, ground so deep you can stick your arm in up to your elbow.

Before the day of automobiles, all the peoples who

traveled through Big Bend routed their trails from water to water. On the way to Oak Spring you can sit in the shade of a Comanche marker tree, a great oak bent in a bow with all its branches growing upright. Comanches marked a good campsite by tying a sapling down; with maturity it naturally assumed a horizontal or bowed position.

As Big Bend opened to ranching, the need for more watering places grew. Ranchers drilled wells, put up windmills, and scraped out stock tanks. Some of these waterholes remain to this day. The wells at Dugout Wells and the Sam Nail Ranch are still maintained. Without regular care such improvements would soon disappear in the desert.

One of man's inadvertent "improvements," the tamarisk or salt cedar, has proved an unwelcome water guzzler. The tree is about the size of an ordinary apple tree, but it loses to the atmosphere about five times as much moisture as an apple tree does. In desert country where water is so scarce, tamarisks pose a serious problem. Brought to this country from the Mediterranean area for use as a windbreak, salt cedar escaped cultivation and spread like wildfire across the Southwest, invading river bottoms, drainage ways, and waterholes in unbelievable numbers. The tamarisk spreads by runners and apparently reaches isolated springs when mammals and birds bring seeds in on their fur and feathers. Growing at the rate of nearly 2 meters (8 feet) in a summer, the deep-rooted tamarisk uses up a disproportionate amount of water and actually lowers the water table. It is useless to man, and wildlife does not browse it because it tastes so salty.

Big Bend National Park conducts a tamarisk eradication program as a water conservation measure centered about the springs. It is hot, dirty, time-consuming work because tamarisks are almost impossible to kill. No known creature can be used for control, and if you leave so much as a root hair, another tree will grow. You have to saw the tree off and paint the stump with a special approved chemical that does not harm other plants or wildlife and will not contaminate the spring. This effort to save the precious amounts of moisture stored in the Big Bend landscape requires constant vigilance and backbreaking effort.

Big Bend has five seasons—winter, spring, sum-

*Who said the desert's palette must be dull? Desert locusts show vivid greens and yellows.*

# Big Bend Ranching Days

Cattle ranching in the Big Bend began about 1870 when Milton Faver set himself up as 'Don' Milton not far from today's Marfa. He eventually built five spreads, including the region's first sheep ranch. As his headquarters he built a fort at Cibolo (Buffalo) Creek Ranch. The Army gave him a cannon for it and even garrisoned soldiers there under his command. During one difficult period, Indian raids wiped out all Faver's livestock except 40 calves confined in the fort. With superb swapping he rebuilt his herds from the Indians' new largesse. By 1880 more and more ranching was pushed west into the Big Bend by range shortages and overgrazing east of the Pecos

River. Formal leasing and land purchases followed. The much sought-after lands had springs. Fencing soon put an end to the free range policy, but as late as 1890 coopera-tive roundups, branding, and drives were still required to sort out whose stock was whose. Stock was stolen by altering a legitimate brand. This came to an end with the introduction of barbed wire, which changed ranching con-siderably. Most of the grass-lands have never recovered from overgrazing. The above photos (left to right) show branding, a cattle drive, and longhorns. Below, a roped yearling submits to inocula-tion.

*Imported from the Mediterranean area for use as windbreaks, tamarisk spread quickly across the Southwest. This water guzzler—it loses five times more water to the atmosphere than an apple tree does—can actually lower the water table.*

mer, fall, and that extra blooming season that bursts out any time you have a good rain and other conditions are right. The more rain, the more spectacular the display, with flowers, buzzing insects, croaking toads, and nesting birds in a complete new cycle of regeneration. Imagine the gravel wastes of the Castolon floodplain awash with flowers—solid carpets of little white and yellow and purple blossoms on either side of the road. Running back from it are desert baileyas and grasses, with orange caltrop blowing like orange butterflies in the wind. Picture Cerro Castellan's red flanks green, with pockets of ochre blossoms amid white heaps of volcanic ash. Imagine Mesa de Anguila's talus slopes misted with grass, Santa Elena Overlook smelling garden sweet and so matted with little low-lying flowers that you cannot put your foot down without crushing dozens. You have never seen their like before and may not soon again, for this is the floral profusion that follows desert rains.

Most Big Bend rains come during the six warm months from May through October, but the expected rainfall may vary greatly with location and with elevation. Thus the super-dry desert between Mariscal Mountain and Castolon averages only 13 centimeters (5 inches) of rain in a year, while the Chisos mountaintops may get more than 50 centimeters (20 inches). Of course some years see more than the average, some years much less. Falling as it mostly does in torrents, very little rain penetrates the thirsty soil. The water just rolls down the slopes, rumbles through mountain canyons, gushes over a pour-off, roars along dry washes, and spreads out over lowland flats in fast-moving sheets heavy with mud. A flash flood can root up and carry off trees and other plants, animals in their burrows, automobiles and their occupants, rocks, and the very earth itself. Then almost as quickly as it came, it may go, leaving gouged and gullied desolation in its wake. Yet in a matter of days, these cracked and peeling mudflats may blossom like a garden.

The reason is that millions and billions of wildflower seeds lie dormant here, waiting for just the right combination of soil temperature and moisture to germinate and burst into bloom. Desert annuals do not store up water as the cactuses do. They do not put down deep taproots as the creosotebush does. Nor do they dress themselves in thorny, waxy, or woody shields as do the desert shrubs. Desert annuals look for all the world like their counterparts in more temperate country. They are just as colorful, just as lavish with leaves, and just as spendthrift with moisture. They live a brief, gaudy life in a hurry, completing the cycle from germination to seed production in a few, short, water-wasting weeks before the desert dries out once again. Then they pass months or even years in the seed stage, waiting for another rain and another burst of luxuriant life. This system works because the flowers produce so many seeds, and because the seeds themselves are marvelously drought resistant and programmed to sprout only at the right time and in the right place.

Each annual and perennial species has its own preferred blooming season and favored locale. The long-legged Big Bend bluebonnet may start flowering in December and keep on blooming until June. Sometimes this rangy relative of the Texas bluebonnet will bloom in such masses that the lowlands look like they have been painted blue. The daisy-shaped nicollet also likes gravelly soils, while the desert verbena does best in disturbed areas. This lavender-pink sweet william, a spring bloomer in the lowlands, appears later at higher elevations as springtime ascends the mountains. As a rule the spring bloom peaks in the lowlands in April, and species that prefer higher elevations flower a little later. Thus the bracted paintbrush begins to flaunt its red flags in June grasslands, and beautiful, deep blue tube-flowers may be seen from May to July on snapdragon vines in the Chisos woodlands. Usually the luxuriance of the spring bloom will depend on the amount of rain that fell during the preceding fall and winter, and the months of June and July are apt to show few flowers at lower, drier elevations. But with summer rains in August and September, many springtime flowers bloom again, sometimes more spectacularly. These rains also produce a bright first flowering from such summer and fall species as the low-lying, sweet-smelling limoncillo, and the broomweed that gold-plates Basin hillsides right through October and November.

Many insects pass the dry months as the seeds do, lying dormant in eggs or cocoons. The same life-giving rains that waken the seeds quicken the in-

# Flowers

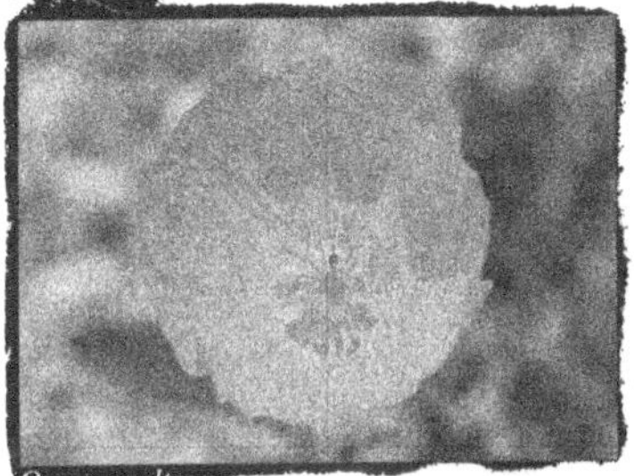

*Orange caltrop*

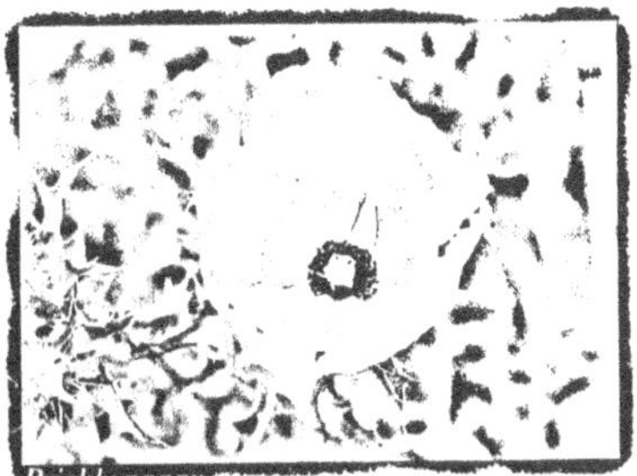

*Prickly poppy*

*Desert baileya*

*Evening primrose*

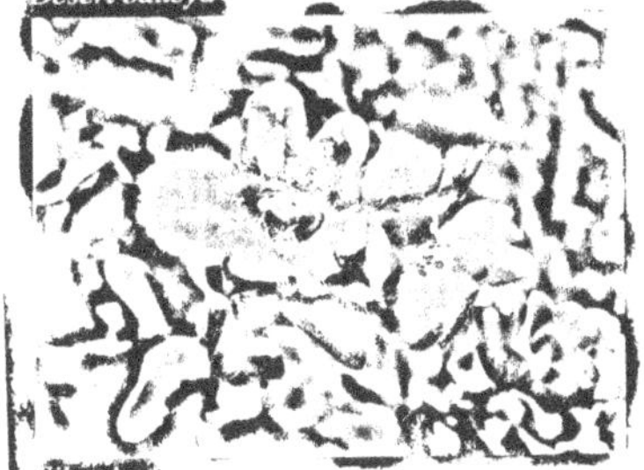

*Silverleaf*

*Dayflower*

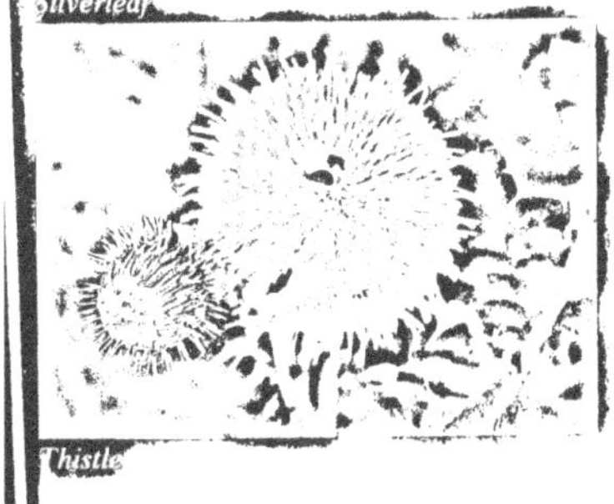

*Thistle*

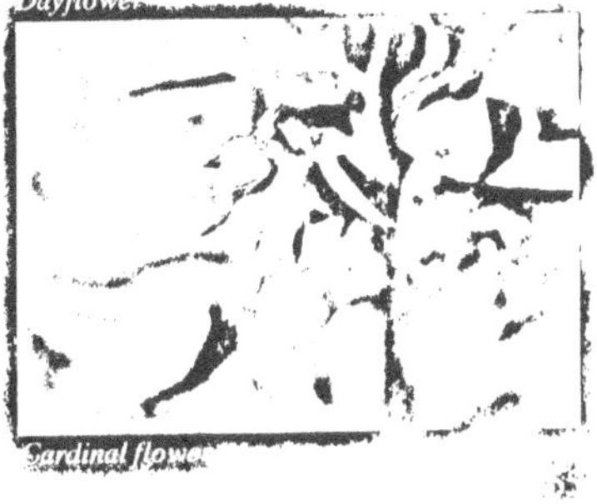

*Cardinal flower*

sects. And as the flowers come into their own there is a mass emergence of flying, crawling, and creeping creatures. The timing of this double emergence is no accident. While the plant-eating insects feed, they also pollinate the flowers.

So beautifully coordinated are these adaptations that specialized flowers attract the very insects that do them the most good. Bees perceive color in the range of the spectrum from yellow through ultraviolet, and you will find them on the many yellow flowers of the pea family, blue larkspur, and lavender ruellias. Bees don't distinguish red and orange, so they pass up the brilliant paintbrush which does attract hosts of butterflies. Flies, beetles, and other insects pollinate relatively unspecialized plants like the sunflowers. And night-flying moths respond to the whites and yellows that almost glow in the dark. In its caterpillar stage, the sphinx moth eats the leaves of the night-blooming evening primrose. When it matures the sphinx moth returns to the primrose and, hovering like a hummingbird, unrolls its retractable tongue and takes up nectar, thus paying its debt to the primrose by pollinating the flowers.

Insect-eating and seed-eating birds capitalize on each rain-induced harvest. The mourning dove, a common park resident, usually nests in the spring, but it may also nest again later in wet years. The scaled quail may produce as many as four broods in wet years, and you may see little brown chicks in mid-October. You may even happen on a young canyon towhee high in the Chisos as late as November. Barn swallows and cliff swallows, both summer residents, nest as soon as they arrive in the spring, and breed again in wet years during August. The black-throated sparrow may nest in both spring and summer if rains have produced a good crop of seeds, while the rufous-crowned sparrow is actually busier nesting in wet summers than in dry springs. The black-chinned sparrow apparently waits for summer's rainy season to nest.

Some mammal populations also rise and fall with the rains. Ord kangaroo rats may not breed at all during long periods of drought, but when a good rainy season produces an abundance of seed, most females soon become pregnant and produce two litters. Females of the first litter may even bear young of their own in the same breeding season.

*Rock-nettle graces the limestone cliffs along the river, one of few plants to do so. You may also find it in the lower mountain canyons. Watch for its blossoms from November through May.*

# As the Wild River Runs

*Ethereal canyon reflections on quiescent waters beckon you toward the timelessness many experience within these vault-like Rio Grande gorges.*

At 3,000 meters (10,000 feet) the golden eagle glides on outspread wings, his head cocked down so he can watch for signs of life upon that motionless desert rolled out like a relief map below him. Largest of Big Bend's airborne predators, the eagle needs an enormous hunting range, and riding the warm air currents high above the border, he can see it all: the flattopped and arched mountains, the sun-bleached lowland, and the silver Rio Grande disappearing and reappearing as it runs downstairs through steep canyons and open valleys. Tilting his two-meter (7-foot) wings, the eagle slipslides for a closer look into a canyon, spots the wake of a surface-swimming snake, folds his wings, and dives like a fighter jet. Before the snake even senses its peril, it is snatched aloft and hangs wriggling in the eagle's talons as the great bird, feathered to the toes, lifts and flies up the canyon with mighty, measured wingbeats.

The waters dripping from the hapless snake's body come from mountains far to the south and north. The Rio Grande begins in springs and snows high in Colorado's Rockies, but backed into reservoirs and doled out to irrigate New Mexico and Texas farmlands, it may hardly even flow below El Paso. What gives the river a new lease on life is the Rio Conchos. This beautiful stream rises in the western Sierra Madres and flows northeastward across Mexico, cutting canyons of its own and joining the Rio Grande at Presidio, 160 river-kilometers (100 river-miles) above the park. Some geologists say it was the Rio Conchos, and not the Rio Grande, that cut those gorgeous canyons in the park. No one knows for sure. But once the river trapped itself, all it could do was dig deeper and deeper by processes that are still at work today.

A walk along a sandbar will show you that the river functions as a practical sorting machine. The water rolling by is so laden with sediment that you cannot even see rocks 13 centimeters (5 inches) below the surface. On the bar itself a layer of curling and flaking mud lies on top of the larger stones and gravel, which have fine sand deposited between them. The heaviest rocks settle out first, then the sand, and finally the finest particles. Water is a powerful lifting and pushing tool, but these waterborne abrasives do much of the river's work, wearing out the rock, undercutting cliffs, deepening and widening the canyons. It goes on at normal stages of water where the river runs less than a meter (2-3 feet) deep, during floods when it crests at more than 6 meters (20 feet), and even during droughts when in many places the river is too shallow to float a boat. So in the slow course of geologic time the mountains are worn away, spread across the valleys, and carried out to sea.

The only streams that have a chance of leaving the desert alive are those whose water sources lie outside the desert. There are few such rivers in the world: the Nile, the Indus, the Tigris and Euphrates, the Colorado, and the Rio Grande. And what a wealth of water-loving life the wild river brings to the Big Bend desert. You can hang big catfish by the gills from your saddlehorn and have your horse walk off with fishtail dragging the ground. So the fishermen tell you.

*Cotton and food crops grew during the first half of this century at Castolon (top) and Rio Grande Village. Both floodplain settlements are popular stops for park travelers today. A third riparian settlement, Hot Springs (bottom), offered resort accommodations in the 1940s.*

Life in this watery world is sustained by a food pyramid based on a super-abundant supply of tiny bottom organisms. A third-meter (1-foot) square of riffle bottom has been found to contain more than 100 organisms. Most are larvae of flying insects: stoneflies, mayflies, dragonflies, damselflies, water and terrestrial bugs, various kinds of flies, midges, and dobsonflies. These curious little creatures have evolved ingenious ways of living, breathing, and eating underwater. Some worm-like caddisfly nymphs build protective cases around themselves, gluing pebbles, bits of shells, and plants together with saliva. They have three pairs of legs up front sticking out of the case and a pair of hooks holding on to it behind, so they can drag their houses with them as they feed. Damselfly larvae breathe through three leaflike gills that project from the hind end of the abdomen, and when warm weather comes they crawl ashore, split their skins, and emerge as gossamer-winged adults. The gills on stonefly larvae extend from the head and thorax, while mayfly nymphs have seven pairs of gills standing out like feathers along the sides of the abdomen. When oxygen is in short supply, mayfly larvae vibrate their gills rapidly so as to quicken the flow of water along their bodies. Some aquatic larvae build nets to catch dinner; a caddisfly nymph may spin a kind of silken windsock that he hangs underwater with the narrow end downstream, using the pressure of the current to keep his prey trapped. Some aquatic larvae eat microscopic plants, some eat insects, and some eat each other. Large dragonfly nymphs may even catch and eat small fish. Larvae are consumed by fishes, frogs, and turtles.

Probably the best way to get to know the river is to get out on it. An easy run is through Hot Springs Canyon, by 90-meter (300-foot) cliffs and over nice little rapids. You put in at the site of the old Hot Springs spa, and take out at Rio Grande Village, having to paddle only at riffles.

Suppose it's early on a fine October morning and you're floating along with the current, watching the sky, clouds, cliffs, and river cane reflecting blue, white, tan, and green on the glossy brown surface of the water. The river is too muddy for you to see what lives in it, but you can see the signs: a spreading circle where a fish has snatched an insect from the surface; mysterious little dimples that look like miniature whirlpools; the beaked head and long neck of a Texas softshell poked up like a periscope. This big turtle's shell is really hard except along the edges, but it is smooth and doesn't have the plates you see on other Rio Grande turtles. The Big Bend slider feeds primarily on plants, while the yellow mud turtle enjoys water insect larvae. And ready to oblige is a cloud of mayflies whirling in mad nuptial flight a meter or two (3-7 feet) above the water. They only live one day and exist as adults simply to mate, but they will sow the river with numberless eggs.

As you round a bend, a pair of great blue herons lifts from the shallows where they've been standing stilt-legged. Now with necks folded and long legs dangling they flap across to the farther shore. Ahead of you a blue-winged teal keeps lifting and settling further downstream. Ducks are seldom seen on the river in summer, but a dozen different species put down as migrants, and some even winter on the river. Now a slim pair of inca doves crosses over-

*A rafter hauls out on a sandbank inside Mariscal Canyon, the middle—and most sheer-walled—of Big Bend's big three canyons. River runners thrill to Tight Squeeze, in Mariscal Canyon, where a rock slab as big as a car compresses the river into a tricky gap.*

head; you see the flash of rufous wings and white tail feathers. On a sandbar stands a spotted sandpiper, head low and tail high. He takes a step, stops, teeters up and down, and then flies.

The sandbar itself snugs in against the cliffside with greenery growing in three distinct tiers. River cane and mature salt cedar stand 4.5 meters (15 feet) tall against the flagstones. And stairstepped in front of these are seepwillows—not a willow at all, but a kind of sunflower that pioneers sandbars—and a younger, shorter stand of salt cedar. The canebrakes fairly crackle with wintering birds: black phoebes, cardinals, brown-headed cowbirds, and a migrating yellow warbler, perhaps. There is plenty for them to eat in that thicket. You yourself discover a nursery of orange true bugs beautifully crossed with olive green, all crowded together in every stage of development on two or three willow leaves. But the most intriguing thing about that sandbar is the record left by its visitors: a lizard's five-toed track with the long unbroken mark made by its tail, and the great blue heron's left-and-right footprints striding along almost in a straight line. You find the cat-like tracks of the ringtail, the dog-like tracks of the gray fox, and the flat-footed print of the hog-nosed skunk, pear-shaped as a bear's. At the water's edge honey bees are collecting moisture to water-cool their hive. One by one they sip and lift off, making a beeline for a cliff.

In the canyons where water flows from wall to wall, you find shore life restricted to those few plants and animals that can make a home on a cliff face. A spindly tamarisk has established a roothold in a thimble-sized deposit of soil just above waterline, and in cracks and crevices higher up, ocotillo and pricklypear are working down from the desert that tops the wall. Empty cliff swallow nests cluster on the undersides of overhangs. In spring you might see baby birds poking their heads from the colony's doorways.

Along the border, people call the Rio Grande by its Mexican name, *Rio Bravo del Norte*. Nowhere does the river seem more wild, more powerful than inside Santa Elena, Mariscal, and Boquillas Canyons. To enter one of these mighty limestone vaults is to understand why mankind has always had to skirt canyon country, and why to this day, except for its historic fords, the river is all but impassable. If you go in by boat the only way out is through. The adventure calls for preparation, knowledge, hardiness, and considerable skill.

But even a landlubber can stand in the canyon's primeval presence. All you must do is make it up the ramps and steps that climb the cliff face at the mouth of Santa Elena, then follow the foot trail down again into the canyon. Looking up from the base of these 450-meter (1,500-foot) walls, you see a vulture and a raven soaring side by side along the canyon's rim. To them you must seem small and as foolishly occupied as the ants drawn up in opposing lines across the sandy path. One step and you could crush the horde. One rock fallen from that height and you are gone. One wild storm upstream and you, the ants, and the sandbar are all washed away forever. Yet you are somehow drawn farther and deeper into the canyon, into this jungle of dark green tamarisk and emerald bermuda grass, through this labyrinth of water-polished boulders, to land's end and water's edge, to the very Beginning that laid these fossil oyster shells in this fierce rock.

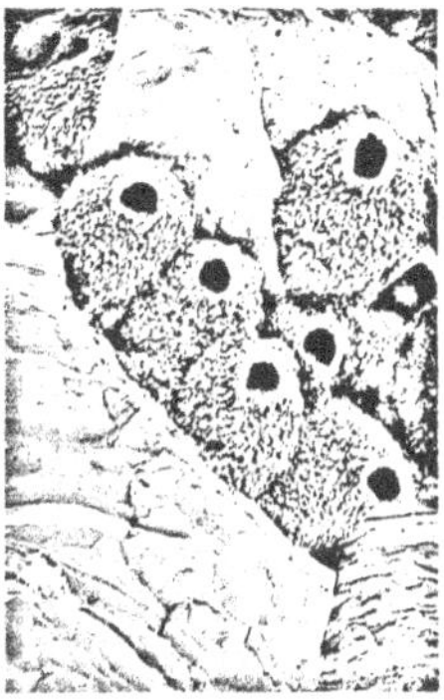

*Cliff swallows colonize with as many as 50 nests in close order. How the adult birds pick out their own condominium from among such clusters remains a mystery.*

Here in the canyon's deep, vault-like isolation the sense of time, that ominous, inhuman distance of the Earth's past, may come over you as the imagined shadow of the wings of a prehistoric reptile, the Pterosaur, perhaps, from 65 million years ago. This was the biggest flying animal ever known to have lived. Picture a 70-kilo (150-pound) flying reptile with a wing spread of up to 11 meters (36 feet), a foolishly long neck, and large head with a long, slender, toothless jaw. Add long legs, and short toes armed with sharp, hooked claws, and a body covered with fur-like material. And figure that each of those long, narrow, glider-type wings was a thin membrane supported by a single overgrown finger, and attached to the body, bat-style, right down to the knee. How could such a huge, ungainly thing ever lift off or fly?

A species does not survive unless it can compete for food and escape its predators, and Pterosaurs, both large and small, existed alongside aggressive, meat-eating dinosaurs for 140 million years. Unlike some smaller species found elsewhere, the Big Bend Pterosaur does not seem to have fished the ocean. At that time, Big Bend offered a river and floodplain environment far from the sea. No one knows how this giant Pterosaur made its living.

## How the Canyons Were Formed

Some 200 million years ago this region lay under a sea whose sediments formed the structural, limestone bedrock patterns of the Big Bend. The basic landscape configurations of today's park were set in motion 75 to 100 million years ago as the landscape emerged, folded, and faulted. Then erosion set in.

The ancestral river that carved Santa Elena, Mariscal, and Boquillas Canyons through bedrock was the Rio Conchos. (The present Rio Conchos contributes most of the water flowing through today's Big Bend. It flows into the Rio Grande just upstream of the park.) When the ancestral river hit the limestone mountain uplifts, it had no alternative but to cut its way through. Steep-walled, narrow canyons resulted. Santa Elena Canyon is cut through the Mesa de Anguila. Mariscal Canyon severs its namesake mountains. Boquillas Canyon, the longest, cuts through the massive limestone Sierra del Carmen. You can see how steep these canyons are by taking a river trip (see page 122) or hiking park trails (see page 119) to the river or to canyon rims.

This region was once much higher in elevation than it is today, but erosion has taken its toll. Mountains and mesas are landscape formations whose rock erodes more slowly than surrounding materials do. Castle-like peaks and high, sharp-rimmed mesas stand as weathered monuments to earlier times when elevations were higher. Such stranded vestiges of

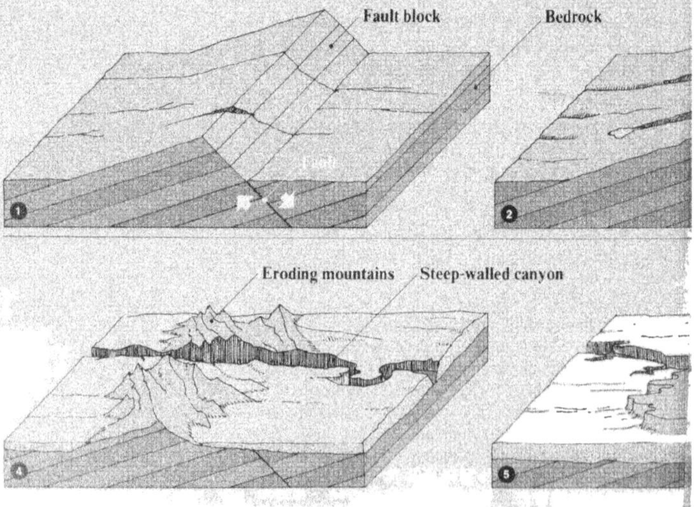

geologic eras punctuate the stark Chihuahuan Desert landscape with eerie architecture. Astronauts have used Big Bend terrain to simulate moonscapes.

The sequence of geologic diagrams shows how the Big Bend canyons formed and what their future would be if slow processes of erosion continue.

❶ Faulting uplifts bedrock to form the mountain mass.

❷ Streams erode the mountains and begin to deposit sediments in the valley.

❸ Streams continue depositing valley sediments eroded from the mountain mass.

❹ Streams have now cut clear through the sharply eroded mountains and formed steep-walled canyons.

❺ All but isolated mesa remnants of the mountains have eroded and weathered away. The riverbed rests in the deep layer of sediments.

❻ In a future stage the bedrock mountain uplift might be entirely eroded away to become a deep layer of sediments.

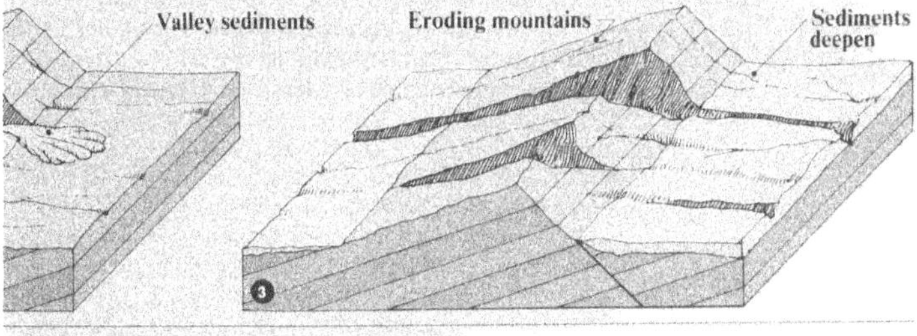


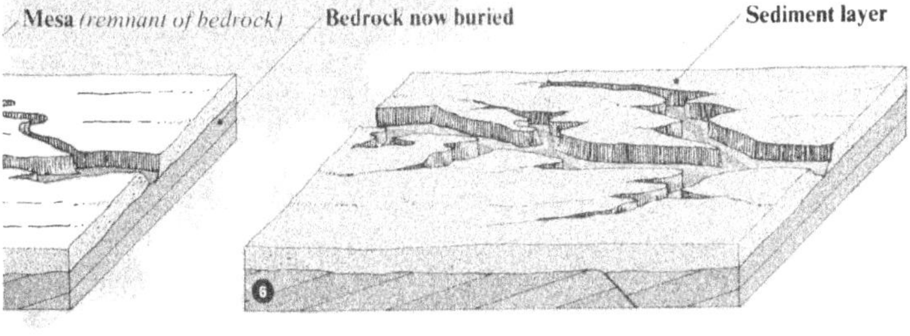

When you think that the Earth is perhaps 4.5 billion years old, that complex organisms have existed for no more than about 700 million years, that the oldest rocks exposed in the park are 300 million years old, and that fossils of backboned animals in the park cover a time span of 70 million years, you can realize how fragmentary the fossil record really is. You get few glimpses of the relatively recent past, but these are astonishing.

For example, geologists know that back in the dim dark distances of Earth time, Big Bend lay repeatedly at the bottom of the sea. Convulsions within the Earth repeatedly raised these sea floors to the tops of mountains, and time after time these mountains wore away. One of the ancient ocean beds can be seen at Persimmon Gap where, in remnants of the park's oldest mountains, fossil sponges, brachiopods, and other simple marine organisms lie exposed. For hundreds of millions of years the three-lobed trilobite was among the most prolific animals in the world, but it had long been extinct when the last great ocean, the so-called Cretaceous sea, washed across Big Bend. Most of Mexico lay submerged and a sort of mid-continent seaway cut North America in two about on the line of the Rocky Mountains. In its early stages this sea harbored ancestral clams, oysters, snails, corals, and a coiled shellfish called an ammonite. You can see these animals preserved in the limestone walls of Santa Elena Canyon. Giant clam shells a meter (3 feet) across and fossil fishes preserved in the round between Boquillas and Mariscal Mountain tell us what lived in later seas. Sea turtles, sharks, and a 9-meter (30-foot) marine lizard that swam in the open ocean have left their remains in the yellowish badlands near the park's western entrance.

Late in this oceanic period the Rocky Mountains began to rise to the north of Big Bend, the Sierra Madre to the south. The park's own Santiago, del Carmen, and Mariscal ranges, and the first upward thrust of the Chisos also occurred at this time. As the mountains rose and started to wear away, delta deposits began to build out farther and farther into the seaway, forming barrier bars and tidal shelves where turtles, snails, oysters, and sharks lived and died. Gradually the near-shore, subtidal environment changed to a tidal flat. This in turn changed to marsh, to beach, to brackish and freshwater lagoons, and finally to an estuary and river floodplain environment. Such was the Big Bend world to which the dinosaurs came. They had been ruling the Earth for eons, but they did not reach the Big Bend area until the Cretaceous sea withdrew.

Sloshing about in the freshwater were amphibious and semi-aquatic species. Probably commonest was the duckbill dinosaur, an enormous reptile that walked on huge hind legs. It had as many as 2,000 teeth. Especially adapted for grubbing up and munching freshwater plants, these flat grinders occurred in batteries in the duckbill's jaw. As a tooth wore out, another popped into place.

Tramping about on all fours and cropping land plants were bizarre looking horned dinosaurs with turtle-like beaks. Armored species included one that looked like a horned lizard the size of a dump truck. Another sported a huge lump of bone on the end of its tail, still another a thick lump of bone above a bean-sized brain. The world's largest crocodile, a 15-meter (50-foot) creature with 15-centimeter (6-inch) teeth, turned up in the park, but remarkably few bones of meat-eating dinosaurs have ever been found.

*Ammonite impressions (top) and fossil clams reveal the era in which the Big Bend area repeatedly lay beneath a shallow sea. Fossil finds show that today's mountaintops were once sea bottoms.*

The cause of the disappearance of the dinosaurs has long been a mystery. Most scientists now believe that a meteor impact about 65 million years ago created a thick cloud of dust that led to the cooling of Earth's climate. Resulting environmental changes may have directly killed the dinosaurs and indirectly allowed mammals to gain an upper paw in the competition for resources. Gradual climatic change and increased competition with mammals may have doomed the dinosaurs regardless of a meteor impact.

When the Age of Mammals began some 64 million years ago, Big Bend lay on an alluvial floodplain where summers were moist and winters were mild. Shallow rivers meandered between natural levees wooded with flowering plants, sycamore, stinking cedar, and tree ferns. Garfish swam in rivers and isolated ponds, while small lakes and swamps lingered on in meander cutoffs and abandoned river channels. It was a land where seasonal floods alternated with dry periods and where, during protracted dry spells, treeless areas invaded the forest.

With such a wealth of habitats a whole new host of animals took over Big Bend. Crocodiles and turtles hung on from the Age of Reptiles, but nature's evolu-

## The Fossil Record

Garfish and turtles in the Rio Grande give us a hint about life here 50 million years ago. Then they swam in waters haunted by crocodiles and visited by the modern horse's earliest ancestor, Hyracotherium; the hippo-like Coryhodon; and Phenacodus, an early species of ungulate related to both hoofed and clawed mammals.

Fossils of these and other animals belie dry and barren Tornillo Flat's earlier Eocene life as a lushly vegetated landscape, as depicted in this reconstruction.

At the time of this scene the dinosaurs ("terrible lizards") had been extinct for 15 million years. Their fossils occur here too. Remains of the giant Pterosaur have been found. These "winged lizards" were flying reptiles whose 11-meter (36-foot) wingspan exceeded that of small jet fighters. The wing was a featherless membrane stretched out from the reptile's body to the tip of its greatly enlarged fourth digit. Fossil remains of Brontosaurus, Allosaurus, Icthyosaur, and others have been found. Fossil Ammonites (see page 67), related to today's sea-

dwelling chambered nautilus, represent the even earlier period when today's Big Bend was covered by a shallow inland sea. Sample fossils are displayed in a shelter off the Marathon entrance road near Tornillo Creek bridge.

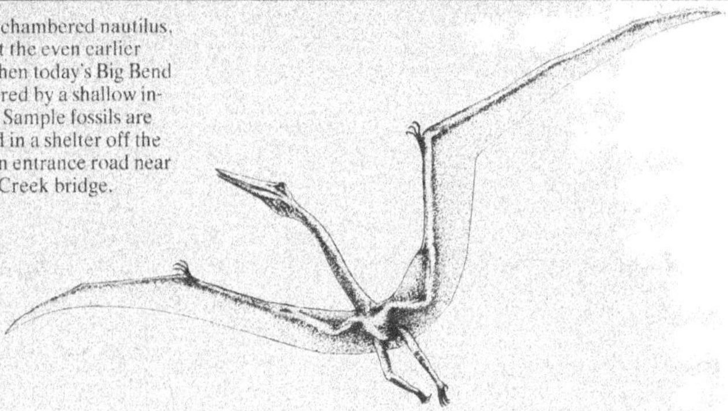

tionary torch passed to the warm-blooded mammals who increased rapidly in numbers, size, and diversity. Remains of 29 species of early, extinct forest-dwelling mammals have been discovered near the Fossil Bone Exhibit site on Tornillo Flat. This same floodplain later accommodated a hippo-like plant-eater, a browsing collie-sized mammal, a panther-like cat, and the little ancestral horse, Eohippus. No bigger than a fox terrier, Eohippus had not yet developed the typical horse hoof and still had four toes on his front feet and three on his hind. He browsed among low forest plants, because nature hadn't yet invented grass.

Some 20 million years ago, when we get our next glimpse of the Big Bend, grasses were well established and the Earth began to witness the rapid rise of grazing animals. In the park, Castolon had a savanna-type environment, with a sub-humid to semi-arid climate. Rabbits and camel- and sheep-like mammals flourished. Plant eaters ranged in size from a tiny mouse to an enormous rhinoceros. This giant was about 3 meters (10 feet) long, stood 2 meters (7 feet) high at the shoulder, and had massive, bony horns on its head. There were also carnivores to fatten on the herbivores.

In the middle of the Age of Mammals, Big Bend country became the seat of widespread and repeated volcanic disturbances, with lava flows, ash falls, and mountains bulging up like blisters as they filled with molten rock. Here, most of these events centered on the Chisos, where the signs can be seen in mountain peaks to this day. Much of what happened since has not even left a shadow; the record and the rocks have both been erased by millions of years of weathering and erosion. But you can see "living fossils" in the high Chisos canyons. These are the ponderosa pines, Arizona cypress, and Douglas-fir trees descended from the moist woodland species that populated this region during the last Ice Age. Occasional remains of the great Ice Age mammoths have also come from gravels in deeply eroded ravines. And you can see all around you the evidence of two latecomers who seem to have reached Big Bend around the same time: Man and the Desert. But here, deep in the river's canyon, both are apparently as remote from you as your own daily world.

*University of Texas paleontologist Wann Langston excavates the sacrum of an extinct sauropod dinosaur. These bones were excavated near Tornillo Creek.*

# Along the Greenbelt and Among the Grasses

*This huge cottonwood tree casts its shade along Terlingua Creek. Early settlers of Big Bend greenbelts used the cottonwood for roofbeams. Later, the trunks shored up mine shafts.*

Even in the dead of night it smells green beside the beaver pond at Rio Grande Village. Well, *dusty* green perhaps, but redolent with reeds and shrubs, trees and grass, with the very jungle breath of the floodplain. On the other side of the river a lone cock crows and close at hand there's a rustling of leaves, a crackling of reeds, the lap-lap of some animal drinking. Leopard frogs croak on in unconcerned bass and baritone burps, sounding like someone's stomach talking. But all at once something has a frog for dinner and the unwilling meal keeps pumping out shrieks, faster and faster and louder and shriller, to the last breath. Then silence. The frog chorus grumbles on.

In the morning you find the flat-footed tracks, long-fingered as a human hand, of that nocturnal hunter. The raccoon is an omnivore, an opportunist with a taste for whatever it can find: amphibians, shellfish, mesquite beans, acorns, cactus fruits, rodents, garbage. The little masked bandit can in fact make himself a campground pest. His dependence on water keeps him a prisoner of the floodplain, but the river, the springs, the sloughs, and the ponds provide abundance in a narrow green world unrolled like a ribbon across the desert. Sometimes this greenbelt, as it is called, is no wider than a bush; in places it may measure three-quarters of a kilometer (half a mile). And it keeps changing, widening with floods, narrowing in droughts, altering course with the river itself.

Just such a change of channels apparently created the bench of bottomland where Rio Grande Village lies today. People have been camping here for thousands of years, as shown by the many deep mortar holes where meal was ground in the limestone ledges near the pumping station. Certainly the place had much to recommend it: wood and water, abundant game, rich soils, an agreeable winter climate, and beautiful views of river and mountains.

After the white man came to stay in the early 1900s most of the native flora changed. Farmers cleared the bottomland to plant cotton and grains, cut down the lanceleaf cottonwoods for roofbeams, dredged ponds and ditches, and enclosed springs. Nowadays, the plants you see are often exotics brought in from outside to create shade and lawns.

Such are the eastern cottonwoods with their heart-shaped leaves, the evergreen live oaks and smooth sycamores, the sweet-smelling honey locust and eucalyptus trees, and that fine green carpet of Bermuda grass. Like the farmers' field crops, these interlopers cannot flourish without irrigation and sometimes even that is not enough. Dozens of eastern cottonwoods have died in recent years.

A good place to see native plants and animals is along the Rio Grande Village Nature Trail, which starts at the campground and leads to a hill above the river. A walk along this trail is doubly interesting because it shows greenbelt and shrub desert side by side, the two habitats often separated by less than a vertical or horizontal meter. The jungle starts at the edge of the clearing and almost at once you come upon a warm spring run. This may be where the rare little Big Bend mosquitofish, *Gambusia gaigei*, originated, but all you see today is a larger relative, *Gambusia affinis*. These five-centimeter (two-inch) predators, little as they are, soon take over. They apparently are invaders from the Rio Grande.

In wet weather this whole spring area becomes a swamp and even in dry times moisture-loving plants crowd the trail. You pick your way beneath black willow trees and grapevines festooned with wild grapes, stop to admire the yellow tube-flowers on the tree tobacco, stoop to avoid a spear slanting from the solid wall of common and giant cane. These two tall woody grasses stand 4.5 meters (15 feet) high and put out great, plume-like flowering heads that shine silver in the sun. Indians used the stems for arrow shafts and ate the roots raw, roasted, or boiled. Many an early settler roofed and even walled his house with cane. Today the reeds supply deer and cattle with attractive browse, and the thickets serve both as home and hunting ground for birds, small mammals, and reptiles.

Here is found the cleverly constructed nest of a wood rat. And there, crouched beneath a bush like some storybook monster is an enormous spiny lizard. Rosy and gray, he peers back unblinking, still as a stone, then melts away so quickly he is gone before you see him move.

A little further on the trail climbs a short rise, topping out on Chihuahuan Desert complete with creosotebush, dog cholla, ocotillo, and the standard collection of cactuses. If you take the trail in spring or summer, the mound-building strawberry cactus will be covered with hot pink blossoms and delicious fruit. By fall the fiercely barbed blades of the false-agave, or hechtia, will have taken on a reddish tinge. In wintertime the tasajillo cactus will be dressed like a tiny Christmas tree in long green spines and bright red fruits. But at any season this water-starved scene stands in sharp contrast to the marsh, rank with reeds and bulrushes, that lies just below.

Unlikely as it may seem you are looking at a beaver pond backed up behind an actual beaver dam hidden in the reeds. The wary, nocturnal beaver is seldom seen and you will look in vain for the familiar beaver lodge, because Big Bend beavers make do with what the floodplain has to offer. Instead of building wooden houses, these rodents dig burrows in the river bank, foraying from there to feed on willow, cottonwood, seepwillow, and river cane. Around the turn of the century the Rio Grande and its main tributaries abounded with beaver, but fur traders trapped them to the brink of extinction, woodcutters and farmers destroyed their food supply, and cattle feeding in the canebrakes trampled their burrows.

It is hard to imagine just what destroying the trees and ground cover can do to fertile land and to a living stream, but a visit to Terlingua Abajo shows you. James B. Gillett, foreman of the famous G-4 Ranch, recalled that in 1885 "the Terlingua was a bold running stream, studded with cottonwood timber and was alive with beaver." There was one grove of trees where he had seen at least 1,000 head of cattle enjoying the shade. But after the Terlingua mines opened, Mexican farmers established a community at Terlingua Abajo, a few kilometers up the creek from its junction with the Rio Grande. All the cottonwoods up and down the creek fell to construction and hungry mine furnaces, and the virgin earth turned bottom up beneath the plow and grubbing hoe.

## The Rare Big Bend Mosquitofish

The Big Bend mosquitofish (*Gambusia gaigei*) has a minuscule geographic range. Not only is it restricted to the park, as some other species are, but it is also restricted to three ponds. The fish was first identified in 1928 in a marshy cattail slough fed by springs located close to the Rio Grande, across from the Mexican Village of Boquillas. In 1954 more were found near Rio Grande Village. This group was later threatened, and a pond was built especially for the fish. But people dumped into the pond other fish that ate this tiny mosquitofish. At one point the world's only survivors were two males and a female, Rio Grande Villagers that biologists had removed to a laboratory aquarium at the University of Texas at Austin. A cold winter again killed nearly the whole park population, and again the Austin-raised stock replenished it. *Gambusia gaigei* gives birth to live offspring and has been around as a species since the mastodons. They feed largely on mosquito larvae.

**Actual size**

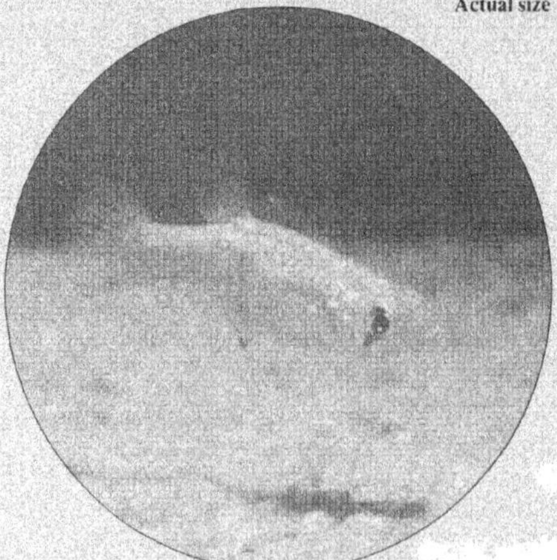

Since the establishment of the park in 1944, some of the more heavily impacted areas have shown signs of healing. Cottonwoods are reappearing along the river floodplain and in Terlingua and Cottonwood Creeks. In some cases, trees were intentionally planted by the park staff to provide shade at campgrounds. For the most part though, the return of cottonwoods to isolated springs and dry washes can be attributed to a natural regeneration process; surviving isolated trees propagate millions of cottony seeds and loose them to the wind. Falling to ground, some seeds take root and, over time and with protection, grow to become mature trees. Each generation in its turn continues the slow and steady healing of the land.

Along the River Road from Castolon to Santa Elena Canyon, you get another glimpse of what water means in the desert. The road winds along the edge of an alluvial bench from 15 to 30 meters (50-100 feet) high. Atop this terrace you have the widely spaced shrubs of the Chihuahuan Desert, but peer over the edge and you find a dense green wilderness crowded between the benchface and the river. All that separates lifeless sand from green jungle is the vertical distance that stops the water. Down there salt cedars have packed themselves thickly into the narrow space.

Another water-loving plant that is prevalent along the river and dry washes is the honey mesquite, a feathery, thorned shrub- or tree-sized member of the pea family. Its root system, with a deep tap root, is so extensive that more wood lies underground than shows above the ground, and natives say you must "dig for wood" here. Mesquite trees multiply as persistently as the salt cedar, but they are highly useful to humans and to wildlife. Honey bees and butterflies visit the yellow mesquite flowers, quail roost in the branches, and wood rats collect mesquite beans in tremendous numbers. Deer, horses, and cattle also relish the beans, while Indians and the early settlers made a nutritious bread from the mesquite bean flour. Honey mesquite carries the standard Kentucky-wonder type of bean. But the screwbean mesquite or *tornillo*—for which Tornillo Creek was named—puts out a curious cluster of corkscrew-shaped fruit pods. Mesquites often mass together in thickets, and they live

*Preceding pages: A greenbelt along a wash contrasts sharply with surrounding desert habitat.*

very well in the drier soils of arroyos, often rubbing thorns with acacias, a family of sweet-smelling, flowering trees and shrubs attractive to honey bees and hummingbirds. The roots of these thickets provide apartments for scores of floodplain pocket gophers, kangaroo rats, and hispid cotton rats.

Despite the luxuriance and rapid growth of floodplain flora, life along the greenbelt rests in precarious balance. The floodplain is the one place in the park from which domestic cattle have never been successfully removed. The problem is that Mexican cows, horses, and burros do not recognize international boundaries.

There was a day when cows grazed grasslands at higher elevations, and before the cows could be seen herds of pronghorns, golden, graceful, large-eyed, with black horns and white rump patches. Imagine if you can how Tornillo Flat must have looked at the turn of the century with fine grasses bending under the wind and the pronghorns flashing their rump patches in sudden semaphore. Depending on speedy flight for protection, pronghorns seek wide-open places where they can see a long way off. Today, Tornillo Flat is a bald shrub desert. The pronghorns vanished long ago. The small band sometimes seen north of Tornillo Creek is the remnant of a herd reintroduced in the 1940s.

The grass that flourished on Tornillo Flat in its heyday was a very good clump-forming grass called tobosa. To the first settlers it looked as though there was more grass along Tornillo Creek than could ever be eaten off. Yet the tobosa soon disappeared, falling first to mowing machines and finally to the domestic herds that cropped it out of existence. Hay balers plied Tornillo Flat into the 1930s.

Grasses spread through seed germination, and in some cases by underground and aboveground stems. As a general rule, root depth equals the height of the grass, but half of the root system dies back each year. This is natural and even desirable since the dead roots create new soil and you still have a sound root system if the grass isn't topped too much. But heavy overgrazing causes grass roots to die back even more and creates an abnormally shallow root system. This is doubly dangerous in desert country where heat and the need to search out water pose crucial problems for all plants. Nor is this all, because other good things come—and other good things go—with a healthy ground cover. As each year's grasses die back they fertilize the soil and insulate seedlings and roots from the sun's killing heat, but when this cover is reduced, good grasses disappear and poorer grasses take over. Surface conditions are finally so changed that neither grasses nor grassland forbs and shrubs can grow. Seedlings may burst into brief life with a rain but when the sun bears down again day after day the unprotected plants perish. With nothing to hold it the soft topsoil blows and washes away. Desert downpours soon gully the ground, changing drainage patterns and lowering the water table beyond the reach of short grass roots. Longer rooted desert shrubs move in, first the shortlived tarbush, and finally creosotebush. The result is seen on Tornillo Flat today: typical Chihuahuan shrub desert with sparsely scattered creosotebush and cactuses.

What happened at Tornillo Flat also happened to other park grasslands. Chino grama, ignored by cattle and goats, but relished by sheep and horses, is and was the dominant grass of the Big Bend foothills. By 1944, overuse had so stripped the grasslands near Government Spring that you could hardly find a bunch of chino on the bare beige hills. Even in the mountains proper the grasses had all but disappeared, and biologists conducting an ecological survey found that the South Rim looked and smelled like a goat paddock.

Once the park was established and domestic animals were removed, native grasses got the chance to reestablish themselves. At first nature itself blocked recovery. For seven long years drought ruled all life in Texas. When the rains finally did come, grasses improved at different rates in different places. They had the most difficult time at lower elevations where heat and aridity are greatest. Several attempts to seed Tornillo Flat have been made, but it may take decades for these once bountiful bottomlands to green again. Grasses in the grassland belt surrounding the Chisos have made a remarkable comeback, however. In ten short years ground cover increased 30 percent. Grasses at the higher elevations may now be as rich as ever.

Different kinds of grasses have adapted themselves to different soils and altitudes and grow in

## The Fur Trade

Gray fox

Mule deer

Bobcat

Furs were an active commodity in the Big Bend until about 1940. Beaver once abounded along the river and its tributaries. Both fur bearers and predators were taken. The bighorn sheep was largely extirpated, selling for a time as "Mexican goat." Wolves were extirpated before the park could provide refuge. Not all species were trapped. The javelina was hunted during World War I, for use in gloves, coats, and suitcases. The bristles went into brushes. Elmo Johnson (photo) bought furs from trappers at his trading post. Good profits required a good eye for raw furs, which Johnson possessed. Furs not trapped during the animal's winter prime were nearly

Kit fox

Coyote

Badger

worthless, and values depended on their condition generally. No fur market existed in Mexico, so furs trapped there were sold at Texas trading posts, destined for the fur houses in St. Louis, Missouri. Only a portion of Johnson's fur stocks shows in this 1929 photograph. Furs were a good business for him, because trappers took their value in goods, his first profit in the transaction. He then sold the furs for his second profit. Later, salaried government trappers worked this area to control predators. Today, all wildlife in the park is protected.

*The U.S. Boundary Survey trekking the Big Bend country in 1852 encountered this Lipan Apache warrior. His portrait came to adorn the survey report.*

company with different co-dominant plants. In colorful sequence the grasses create a three-dimensional mosaic that begins on the edges of the shrub desert, covering the foothills and running up through the woodlands out onto mountaintop meadows. You can see these differences quite easily.

Let's say it's ten o'clock on a fine fall morning, and you are driving from Rio Grande Village up to the Basin. You find shrub desert all around you at first, with the usual creosotebush, ocotillo, and pricklypear. The only grass you see grows in a buff-colored strip on either side of the road, where runoff from infrequent rains creates a habitat moist enough for grasses to germinate and reach maturity. But after you pass 900 meters (3,000 feet) of elevation you begin to see the first sparse bunches of chino grama spotted here and there across the desert. Somewhere just below the Dugout Wells turnoff you find quite a bit of bunch grass among the bushes. Almost imperceptibly the chino thickens until at 1,000 meters (3,500 feet) the hillsides appear cobbled over with lumps the color of gray rock. It appears that you couldn't step down without stubbing your toe against a clump of gray-green grass or lechuguilla. Soon all the slopes are patterned yellow and gray-green as solid stands of lechuguilla crowd into the chino grama. The grass is so thick now that scaled quail hunt seeds along the roadside and cottontail rabbits bounce jauntily across in the mid-morning sunlight.

The colors you see owe much to the angle of the sun, but at about 1,200 meters (4,000 feet), near Panther Junction, the plants themselves add a soft new hue, the beautiful silver-blue of ceniza shrubs. Bunches of vivid green grass crop up amid the tawny three-awns by the roadside, and the tops of the tarbush are brushed with yellow. Rock, gravel, grass, and shrub blend together in shades of tan, gray, green, and blue pinpointed by Christmas-red fruits on the guayacan, with once again vast gardens of chartreuse lechuguilla and skeleton-leaf goldeneye shrubs. These rolling grasslands have much less pricklypear and creosotebush, and they have a lower profile than the shrub desert. Here and there the Torrey yucca lifts its shaggy head high, but the whole land surface looks somehow smoother, less hobnailed than before.

When you start the 11-kilometer (7-mile) climb

toward the Basin, chino grama gives way to a variety of grasses, blue, black, hairy, and side oats grama, and the straw-colored grass aptly named tanglehead. The whole color scheme changes with the grass from shades of gray to reddish brown and yellow, and the tops of these good grasses have the heavy headed look of grain. The lechuguilla becomes less obvious. Gradually the moisture-loving mountain shrubs push the grasses back from the roadside, but the golden gramas grow with wonderful abundance on the drier slopes beyond. Still higher, the still good grasslands are studded now with century plant, basket grass, and sotol.

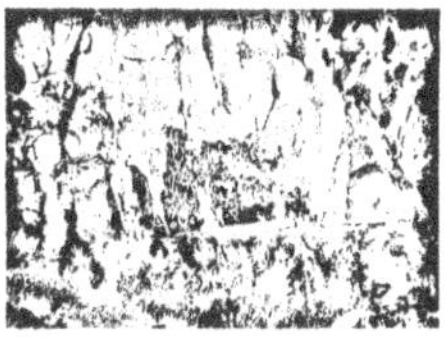

*Ancient rock art known as petroglyphs (top) decorate rock faces. The arid climate protects them. Mortar holes (bottom) deepened imperceptibly each time the Indians ground seeds or grains in them.*

Earlier in this century sotol abounded between Green Gulch and Government Spring, but in the severe drought during World War I, ranchers chopped it out for cattle feed. Today you will find a perfect forest of sotol growing at Sotol Vista on the western flank of the Chisos. As you can see from the overlook, this narrow-bladed plant favors the cooler north-facing exposures. Sotol leaves grow at ground level from a short trunk, and nearly every spring the plant thrusts up a brush-like bloomstalk, 4 to 6 meters (15-20 feet) tall, covered half its length by greenish-white flowers. Prehistoric Indians had many uses for sotol, and later the Chizos often located their rancherias at solid stands of "desert candle." They used the fibrous leaves in making mats and ropes, fermented a potent drink from the pineapple-like heart, or roasted it in rock-lined pits as we roast beef. Several crumbling sotol pits are found throughout the Chisos, and sotol still serves as an all-purpose plant along the border.

Sotol is one of those Big Bend plants that has its very own grasshopper. If you look closely, you may find a little fellow feeding high on the bloomstalk and looking for all the world like a bud. He lives his whole life on the sotol plant and knows just how to use it to advantage. If you annoy him he will probably play possum, drawing in his legs and dropping like a stone right down into the thick of the sotol leaves. Few predators come away unscathed from a close encounter with those saw-edged blades.

Another lily of the desert is the yucca. Overabundance of this perennial often indicates abuse of a grassland, and yucca gradually thins out again once a good grass cover is reestablished. Several species

## Indians of the Big Bend

Evidence of Paleo-Indian culture dates back to about 9000 B.C. here. The Archaic and Late Pre-Historic culture sites date from about 6000 B.C. These occur near today's water sources, so climate conditions then were probably similar to now. The Archaic people used the atl-atl, a dart-pointed throwing stick, to hunt game animals. The Late Pre-Historics used the bow and arrow. Spaniard Cabeza de Vaca encountered their descendants in his 1535 expedition through the area. In the 1640s the major Indians here were the Tobosas, Salineros, Chisos, and Tepehuanes, who fought Spanish encroachment and enslavement. Spanish horses enabled the Mescalero Apaches to expand their range and dominate the area by the 1740s. By the 1840s, the Comanches, also with Spanish horses, dominated an enormous range focused on the Big Bend. When Texas was annexed by the United States in 1845, the U.S. military became the Indians' antagonist. Forts strung along the route to California goldfields after 1849 sliced the Indian territory in half.

Retreating far into the mountains, the last Apaches, under Chief Victorio (left inset), were defeated by Col. Benjamin H.Grierson and scattered into Mexico. There Victorio was soon killed by Mexican

troops. Col. William H. Shafter (right inset) continued the Army's Indian pacification after March 1881. Not all Indians were hostile. The painting, by Capt. Arthur Lee, shows West Texas Indians trading at Fort Davis.

of yucca flourish in Big Bend. Most widely scattered is the Torrey yucca or Spanish dagger. Seen from desert lowland to mountain top, it thrives at Persimmon Gap near the park's north entrance.

Yucca grows like sotol from a central trunk, but it adds new growth at the top and lets its dead leaves drop down in a palm-like grass skirt. The yucca also puts up a bloomstalk year after year, and the creamy flowers are a favorite browse of deer and cattle. Yuccas may live from 50 to 75 years, reaching tree-like proportions. The giant dagger that gives Dagger Flat its name may grow more than 6 meters (20 feet) tall. In good years it blooms around Eastertime, and a single bloomstalk may have more than 1,000 flowers and weigh as much as 30 kilos (70 pounds). Indians used nearly every part of the yucca in countless ways. They ate the flowers and fruit, wove baskets and made brushes from leaf fibers, and made soap from the roots. Along the border, natives still harvest the great yucca flower-heads and feed them to livestock.

*One of six yucca species in the park, the giant dagger abounds in Dagger Flat. It blooms every two or three years from late March through April.*

The Mescalero Apaches roasted the mescal or century plant using the same technique they used for sotol. The mescal was as central to the Apache way of life as the bison was to the plains Indians. They made food, drink, medicine, mats, ropes, bags, and even needles and twine from the great, gray-green agave.

It is not true that the century plant takes a hundred years to bloom; it's more like 25 to 50 years. Once started the bloomstalk shoots up at the rate of nearly 2.5 centimeters (an inch) in an hour, or up to 41 centimeters (16 inches) in 24 hours. Soon platters of golden flowers float on the air, and bees, flies, and ants come to the feast. The violet-throated Lucifer hummingbird may sometimes be found at a mescal in blooming season. At night the Mexican long-nosed bat leaves its cave on Emory Peak and comes to feed on the nectar and ample pollen.

Like all agaves, the mescal expends its strength on this onetime burst of blooming. Once the seed pods form, the plant dies and young plants arise near its base. But that is not the end of its beauty or its usefulness. If you climb the foot trail to Juniper Flat through the thickly grassed woodlands above the Basin, you will pass more than one dead mescal with its towering bloomstalk supported by the limbs of some pinyon pine. Year by year, in graceful and gradual decay, the agave leans in its neighbor's embrace. As it slowly splits up, it dresses the pine in a flowing drape of fiber fragments. You can see little wormholes here and there, and larger holes where woodpeckers have taken a meal. This life-supporting return to dust goes on all over these mountains.

Many of the park's finest grasses may be found in the mountains, for while the grasslands blend into the woodlands somewhere around 1,700 meters (5,500 feet), the grasses continue all the way up to the mountaintops. Again, the species change with climate, altitude, and ground cover conditions. The beautiful bull muhly favors north-facing slopes, and, growing in clumps up to just over a meter (4 feet) high, it helps hold soil on the hillsides. With its large purple flowering heads, it makes a good bedding grass for deer. So does the golden thread-leaf stipa that likes to grow under juniper trees, spilling its long blades in rounded clumps.

Since the return of the grasses the mountains and foothills both support a fascinating host of grassland animals. Some, like the yellownosed cotton rat, have come back with the grass. Many of these grassland creatures are abroad at night and are best seen at dawn and dusk. Even after dark you may see a great deal of life along the highway. The 16-kilometer (10-mile) stretch between park headquarters and the Basin may bring startling glimpses of grassland activity within reach of your headlights: a great horned owl with its catlike face and "ears," sitting still as a sphinx in the middle of the road; a coyote pair gamboling like shepherd pups, their paired eyes flashing white and bright as automobile headlights. These carnivores have come to hunt the rats and rabbits making a mad dash across the road. Arthropods and snakes come to the pavement to warm themselves. The diamond sparkles that litter the blacktop may be reflections from spiders' eyes. You see a blood-red coal of fire spring to one side at the edge of the road and then as abruptly as a UFO leap 3 meters (10 feet) straight-up into the air. A ringtail climbing a tree perhaps? Or a poor-will taking flight? A flash of pink lights at elbow and shoulder height turns out to be a mule deer and fawn. Stop, turn off your own lights, and look and listen and you will behold a world as it must have been a long time ago, singing in the silence underneath the stars.

*The century plant took its name from the erroneous notion that it took 100 years to bloom. It takes "only" 25 to 50 years. Then the plant dies.*

## Bandits and Revolutionaries

When the deepest channel of the Rio Grande became the boundary between the United States and Mexico in 1848, the Army assumed a thankless task: To enforce an invisible boundary in impossible terrain populated—if at all—by people traditionally disposed to crossing the river freely. At times the Army depended on the Texas Rangers. During the Civil War era troops were withdrawn from the Big Bend, and border incidents surged. After the Mexican revolution of 1910 that government lost control of its northern provinces. Bandit gangs plagued both Texas and California. Pancho Villa used border forays to create tension between Mexico and the United States. In 1913, raiding ban-

dit Chico Cano was captured and freed by his gang in an ambush; the next year he killed his Customs Service captor. Cano was photographed (right) with 8th Cavalry Major Roy J. Considine in 1918. Three soldiers and a small boy were killed in a 1916 raid on Glenn Springs led by Navidad Alvarez, a lieutenant of Pancho Villa. The store was looted and houses burned. Livestock raids on ranches sometimes left entire families murdered. Many Texas Rangers were killed in ambush over the years. Smuggling activities abounded. Arms and ammunitions passed into Mexico, and liquor (during prohibition) and silver bullion passed into the United States.

# Where Mountains Float in the Air

*The Chisos float in morning mist after a rare ice storm.*

From a distance the Chisos Mountains don't look wooded. They seem to be sculpted from naked rock. Even as you begin the drive up Green Gulch you expect to see only buffy grass, century plants, and sotol. But as you pull abreast of Lost Mine Peak the grass is gone from the road's edge and the shoulders are covered more and more thickly with green, leafy shrubs. Soon you see taller bushes on the open slopes, the evergreen sumac, Texas madrone, cat claw, and common bee-balm, crowned with beautiful deciduous and evergreen leaves.

The Chisos Mountains fascinate for so many reasons: their great beauty, their wooded coolness and greenness, the exceptional variety—and rarity—of plant and animal life. Lying close to the border, they encompass an interesting blend of U.S. and Mexican species. But because they stand alone and apart, they support species found nowhere else. And their height creates a temperate environment, a grassed and timbered island in a sea of rock and sand.

In the Chisos as elsewhere in desert country, what lives where depends on highly local conditions of soil and climate. While daytime temperatures in the mountains average significantly cooler than the desert flats, and while rainfall in the Basin is twice as heavy as at Rio Grande Village, just as crucial are the land's shape and the sun's angle. You might expect that trees grow as a woodland belt around the mountains, with treeline determined by altitude. Not so. Many trees prefer canyons to exposed slopes because soils are deeper there and drainages offer more water. Even in the same canyon, more trees may grow on cooler north-facing slopes, while grasses predominate on sunny south-facing slopes. Or pinyons, junipers, and oaks may flourish at lower elevations, while most other forest species occupy the canyon's heights.

At about 1,400 meters (4,500 feet) of elevation as the mountains begin to close, a surprise delights the eye: The first of what you might call trees beside the road. Suddenly, masses of trees—junipers, little oak trees, and pinyon pines—higher up in the drainages that crease the mountainside. And multitudes of woodland birds dart in and out of the trees and bushes along the roadway. You understand now why they call this valley Green Gulch. As you approach the first water tank you are in actual woodland. Grasses still grow on the open slopes along with bushes and pricklypear and sotol, but here are the first easily distinguished little pinyons beside the road. At 1,600 meters (1 mile) of elevation they are fairly respectable pines. Tree cover increases and, after the second water barrels, the woodland comes right down to the road. The grass is less obvious now, but it covers the ground between the trees and is especially thick on south-facing slopes. Flocks of little birds fly up from the road ahead of you. At the switchbacks in Panther Pass you peer up the steep slopes at solid stands of deciduous green trees set off against darker evergreens. Way up at the top leap scarlet flames of frost-touched Grave's oak and fragrant sumac.

These pinyon-juniper-oak woodlands are pretty typical of the dry Southwest. Chisos woodlands include only one pine, three junipers, and many varieties of oak. You can easily identify pinyon and juniper trees, but the oaks are often hard to tell apart because they tend to hybridize. The different tree species have different soil and water requirements and where moisture is harder to come by they grow smaller and more widely spaced. You'll find redberry juniper at drier, lower elevations, such as Green Gulch. This is a rather scraggly shrub or small tree whose lower branches often touch the ground. It has scale-like yellow-green leaves, and red berry-like cones. It will often invade abused grasslands. The gray oak, an evergreen with small olive-green leaves and dark gray bark, also prefers drier soils. On exposed slopes it too takes on shrub-like proportions but grows to 20 meters (65 feet) in protected canyons. Birds, peccaries, and deer feed on gray oak acorns. The Indians preferred the acorns of the Emory oak.

The other narrowleaf evergreens prefer intermediate to moister soils and the Chisos Basin is a good place to see both drooping and alligator junipers, the latter named for the square scales that make its bark

*Startling plant combinations comprise this forest floor on the South Rim of the Chisos. Such biotic richness and surprise led to the park's designation as a Biosphere Reserve.*

*Following pages: Volcanic spires remind us that change has not always come slowly in the Big Bend. Molten rock under intense pressure created these spires as plugs inside softer rock, which has long since eroded away.*

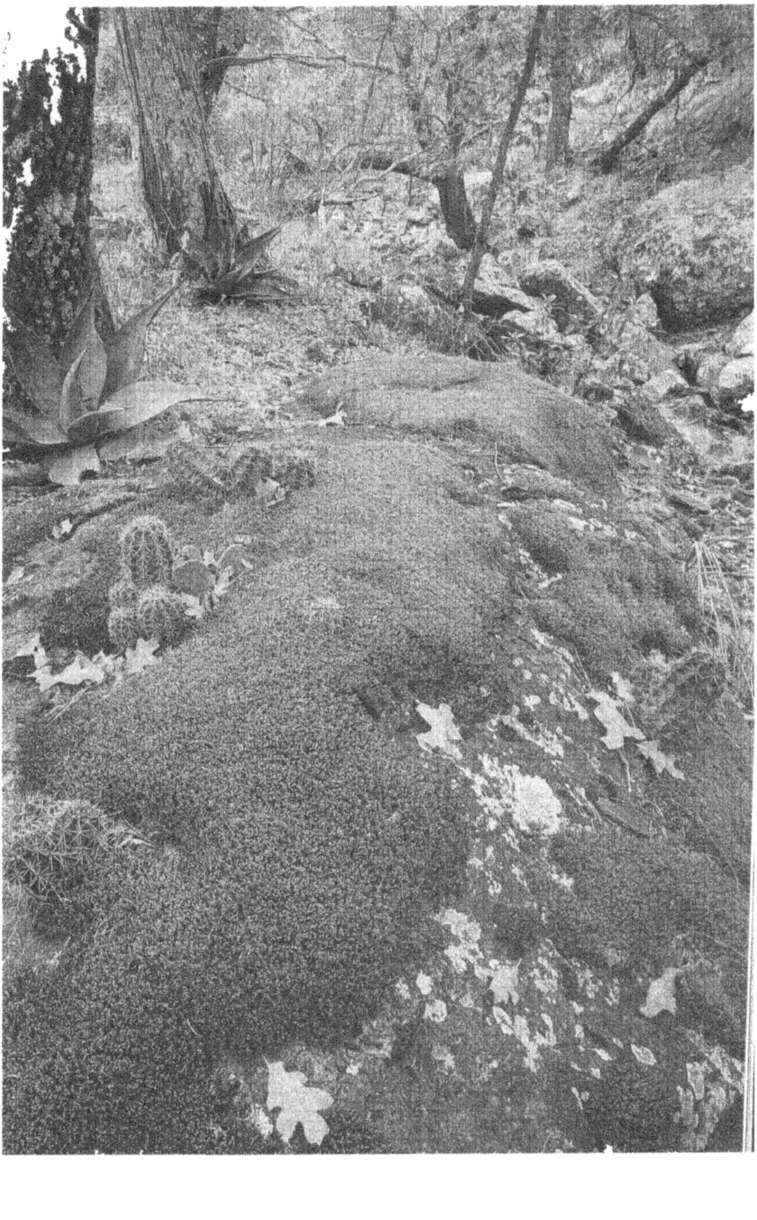

look like alligator hide. This slow-growing, long-lived tree has bluish-green needles, and gray fox and rock squirrel relish its berry-like cones. Drooping juniper all but cries out for recognition. Its wilted leaves and drooping branches seem to be dying of thirst, but in fact are perfectly healthy, as is the bark that shreds in long, fibrous strips. You'll find plenty of drooping juniper in Mexico, but the only place you can see it in the United States is right here in the Chisos Mountains.

The Chisos Basin has been hollowed out of volcanic rock by stream erosion. The peaks that ring the Basin all came into being when molten rock squeezed up under enormous pressure from deep within the Earth. Some of the red hot stuff poured out over the land surface in lava flows which cooled so quickly that they cracked in long vertical fissures. Then as the ages passed, joints toppled and square-faced peaks, buttresses, and free-standing spires emerged. So the Basin wall took shape from Casa Grande southward through Emory Peak. Later on, more molten rock pushed up from below, bulging the surface rocks upward without breaking through. Again the eons passed and the softer surface rocks wore away, exposing the dome-shaped peaks which now rim the Basin to the north and west. Today, loose rocks and clays still inch downhill toward the Window. All Basin runoff heads for this chute, and when it storms in the heights, the dry waterfall turns into a torrent, with boulders bouncing along like so many corks. Growling and grumbling, the big rocks plunge over the pouroff in a 67-meter (220-foot) free fall, coming to Earth in a great rubble pile below.

Pinyon pines grow abundantly across both Mexico and the American Southwest. In the Chisos you'll find them almost anywhere above 1,500 meters (4,800 feet) of elevation, and at lower elevations they will be the only pines. Short of trunk, with spreading lower branches, egg-sized cones, and short, slender, bluish-green needles, pinyons range from dwarf size to a tree 15 meters (50 feet) tall. Many birds and mammals eat its delicious nuts.

As elevation increases you may find fewer junipers, while more pinyons appear on the open slopes and more oaks along drainages. The deciduous Graves oak requires more moisture than other oaks, so you find it putting forth its shiny, dark green leaves in high moist canyons. The Emory oak also prefers high drainages but it grows at slightly lower elevations. It has small lance-shaped leaves. The Chisos oak, a small, graceful tree with narrow, leathery leaves, requires a high water table. In all the world it grows only in the Chisos Mountains' Blue Creek and Juniper Canyons.

Many of these interesting woodland features can be studied at leisure in Upper Green Gulch, reached by the Lost Mine Trail from the trailhead in Panther Pass, following the well-kept path at least as far as Juniper Canyon Overlook. Here you find yourself among the very pines and oaks that you viewed from the switchbacks far below. On location it appears much as it did from far below. There are few grasses and a host of flowering bushes. Shaggy mountain-mahogany and fragrant sumac make excellent feed for whitetail deer. Fragrant ash puts out long clusters of cream-colored flowers in springtime. Mountain sage, a beautiful shrub that grows nowhere else in the world, bursts into crimson flower each fall. As a hummingbird feeder it even outranks the golden platters of the century plant. Probably most surprising at this elevation is the persistence of desert and grassland plants, for here among the pinyon pines and oak trees grow clumps of ocotillo, lechuguilla, pricklypear, and the great gray-green blades and towering bloomstalks of the Big Bend agave.

Dryness is a fact of life in these woodlands. The north-facing slope is densely covered with trees, while the opposite south-facing slope exhibits mostly ocotillo and lechuguilla. The same holds true on the narrow ridge extending from Casa Grande. Pine woodlands face north and lechuguilla flourishes on the south-facing slope. It's a question of solar exposure and resulting temperature and moisture variations.

Oddly enough, you will likely see more wildlife in the populated Basin than along the whole Lost Mine Trail, for animals find the Basin as attractive as humans do. The del Carmen whitetails find it a good place to feed off and on throughout the day. These deer are found only in the Big Bend and across the river in the Sierra del Carmen. To the rock squirrels the Basin offers acorns, pine nuts, and plenty of rocky lodgings. The busy cactus wren can indulge its habit of year-round nest building, because there is abun-

*Mule deer (top) frequent the park's lower, drier elevations. The Sierra del Carmen whitetail (bottom) haunts the Chisos Mountains. It has much smaller ears than the mule deer and bears the characteristic flag tail. Isolation allowed this sub-species of whitetail to develop. These deer live only in the Chisos and across the river in the Sierra del Carmen.*

Coyote

Should the jackrabbit reflect on its situation it might think life is a conspiracy. This prolific breeder might feel it exists solely to keep the local web of life functioning. Many predators include this big-eared, nimble-footed creature in their diets.

Bobcat

Golden eagle

Big ears, big feet, and protective coloration are the jackrabbit's major survival mechanisms. It will listen for danger, sit motionless until perceiving a threat, and then burst into speedy flight. The ears may also represent an adaptation through which the jackrabbit can emit excess body heat to the environment, a useful ploy in deserts. Birds, snakes, and numerous mammals prey on the jackrabbit, which is really a hare, and on the park's desert cottontail and eastern cottontail rabbits. The coyote can sustain bursts of speed sufficient to run them down. An arch opportunist, the coyote eats almost anything, including an occasional tennis shoe. The bobcat and mountain lion (photo page 103) are the park's two felines. The bobcat hunts by stealth. It will sit by a game trail for hours and then pounce on passing prey. Threats to the jackrabbit come from above, too, where the golden eagle soars.

dant tall grass, and introduced yucca, a favorite avian building site. Sounding like a child's squeeze toy, the canyon towhee takes to his human habitat as freely as a house sparrow, and the crestless Mexican jay scolds ferociously, as jays will.

Toward twilight the whitetail deer put in another appearance. These dainty little animals are highly territorial, so you are likely to see the same band of bucks in the upper Basin, and the same doe and twin fawns near the campground turnoff. The eastern cottontails, larger than the desert cottontails, also come out at dusk, after spending the day in the very same thickets that the deer come to browse. A little later, skunks may appear. These spotted, striped, or hognosed nighttime foragers are cyclic in their populations, and like the raccoon they have a decided fondness for campgrounds.

In the Basin as elsewhere, many residents are heard rather than seen, especially the tree crickets and katydids that sing their songs at night. The Chisos Mountains even boast their own katydid, known nowhere else in the world, the Big Bend quonker. Scraping its wings together it produces a squeak much like that of a cork being pulled from a wine bottle.

The woods that rim the Basin to the south have a softer, more life-supporting look than the slopes above Panther Pass. They feature the same junipers and pinyon pines and the same pricklypear and century plant, but the difference is the grass. Tall and short, gold and blue, tasseled, tufted and feathered, it grows so thickly under the trees and between the shrubs that it all but covers the ground. You find basketgrass, too, with leaf edges like fine sawteeth and their tips frayed into curls of twine. With such abundant cover animals should thrive, and judging from the scat they do. In the fall the pricklypear is still in fruit, and it appears that everyone is eating brick-red tunas. At lower elevations where the tunas have already gone by, many animals are eating black persimmons.

But strong as the signs of life are, the evidence of death and dying cannot be ignored. You hear the shrieks of some creature on the edge of a little meadow, watch a hawk come to circle the tree tops, wonder who has won that contest as the cries cease and the hawk flies off. You smell the pungent odor

*The mountain lion's mystique explains the many park place names that bear its alias, panther. Sightings of these regal cats occur throughout the park, especially at dusk or after dark.*

## Quicksilver Mining

Quicksilver, or mercury, the only metal that remains liquid at ordinary temperatures, was mined as cinnabar in the Big Bend country from about 1884 until after World War II.

Cinnabar, red mercuric sulphide, was used as a pigment and medicine as early as the first century. Indians used it as pigment for war paints and pictographs. Today mercury is used in electrical apparatus, control instruments, thermometers, and medical and dental preparations. The United States once produced about one-third of all quicksilver. From 1910 to 1920 Texas mines produced about one-third the U.S. production. Locally, quicksilver mining began in 1884, but real production began after 1896. The park's Mariscal Mine was opened as Lindsey Mine by D.E. Lindsey, an immigration inspector, about 1900. Production increased greatly about 1916, under the ownership of W.K. Ellis, as World War I pushed up

quicksilver prices. The mine floundered again with post-war price declines and was not profitable anew until World War II. In the Mariscal Mine's heyday between 1919 and 1923, from 20 to 40 men worked it. All were Mexicans except the manager, foreman, and a brick-kiln specialist. Wood for the furnaces came from as far as 80 kilometers (50 miles) away, by burro. These photographs show quicksilver operations at the Waldron Mine, just outside today's park, in 1916. Laborers toiled 12-hour shifts for $1.00 then.

*Aplomado falcons once ranged this far north and east, but no longer. Now that the national park offers a large protected area, we hope these birds of prey will return.*

of pine resin and follow it to its source, a pinyon oozing crystal drops from dozens of holes. As the sticky stuff ages it turns yellow, and there, mired in the gum, is a small black beetle exactly as some fossil bee in amber. And here is a redberry juniper so strangled by pink tree-thief that the greenest thing on it is the mistletoe cluster.

The death of trees is actually part of the continuing cycle of life because it returns much needed nutrients to the soil. Termites play an important role in converting dead wood into substances useful to plants although they cannot digest the wood themselves. Tiny protozoans living in their intestines secrete digestive juices that do the job for them. Dead trees also provide nesting and resting places for various birds and mammals: Screech owls and mice may make their homes in hollow trees and logs, and so may the ringtail, gray fox, and bobcat. These four mammals are adept tree climbers. The brush mouse climbs to garner pine nuts, acorns and juniper berries, while the ringtail and gray fox add berries to supplement their largely meat diet. The bobcat, as sure-footed aloft as other cats, will take to the trees when pursued but prefers to hunt on the ground.

The bobcat, so-called because of its short tail, limits his diet to mammals, birds, and insects. This smallest and most common of the wild cats prefers rocky canyons and outcrops in pine-oak woodlands. Hunting mostly at night and on the ground, he prowls on padded feet, hides for hours beside a game trail, and springs on his prey in one lethal pounce. The ringtail is the busiest of small predators. Strictly nocturnal, he covers a wide territory several times each night. He has much the same tastes as the raccoon, but without the latter's fondness for water. The ringtail especially likes to prowl rocky ledges and canyon cliffs on the lookout for insects and small rodents. He climbs easily and hunts in trees for roosting and nesting birds.

Water, those rare spots where it occurs permanently, can do astonishing things to a woodland. In a secluded canyon you'll find a grotto no bigger than a room, with countless seeps trickling down the face of a high, nearly dry waterfall. At the foot of the fall, maidenhair ferns and stream orchids crowd beside a deep pool, while redbud, oaks, and maples canopy a burbling brook. The sun rarely reaches this rock garden, so that at midmorning it is significantly cooler than the grassy slopes nearby. The water in the main pool is even cooler. So many big boulders lie heaped across the canyon that few animals can reach the water, yet all sorts of creatures live here. You'll find leeches in the pool, and, looking like a shelled peapod, a dead katydid's exoskeleton gutted by water insects. You'll hear a canyon wren echoing its own song, and spy tiny canyon tree frogs clinging to trees and rocks with sticky little suction-cup mounted toes. Tiniest are the mites living in parasitic comfort on the tree frogs. Such microenvironments stand in surprising contrast to the grandiose environment of the mountain masses surrounding you here in the Chisos, and elsewhere.

*The peregrine falcon's easy soaring belies its diving speeds of up to 320 kph (200 mph). Peregrines nest in the river canyons and high Chisos.*

People call one particular spot high in the Chisos the top of the world. If you sit there at sunset you can watch the turkey vultures describing long, lazy figure eights from the top of the world down to the Window and back. At that time of day those vultures, not looking for food, simply seem to be enjoying their world and being alive. In those sunset cruises they're living to the limit of their unique life form—and are glorifying Big Bend!

Big Bend's future, as its past, will be ruled in the long run by triumphant nature. The vulture's self-celebration almost portends this. If we are entering another Ice Age, what a new lease on life this will be for the high mountain forests of Douglas-fir, ponderosa pines, Arizona cypress, and quaking aspen. What vindication for the staying power of these many beleaguered species. If on the other hand the desert pushes on, the summer rains don't come, and the springs dry up for good, then Big Bend's big trees will vanish like the dinosaurs.

Such events are beyond our power to influence or foretell. Indeed, change may be so imperceptible, so slow, that people, supposing there are people left in Big Bend, may find it perfectly natural. Or change may be catastrophic, and those last Big Benders disappear without a trace, as though snatched off the Earth. And those, if any, who come after may then marvel over ruins and artifacts and ask what drove these Ancients from their homes.

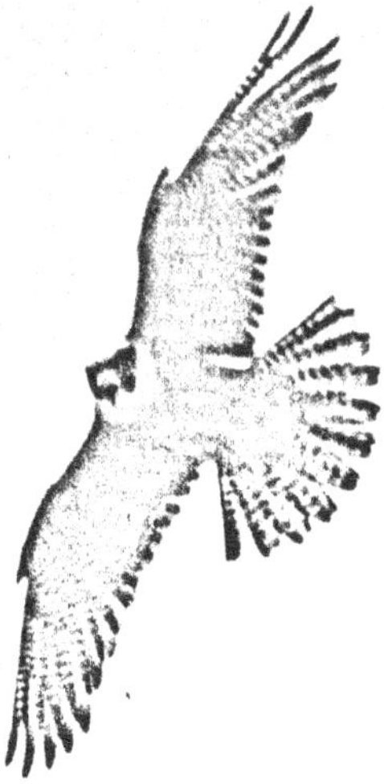

Guide and Adviser

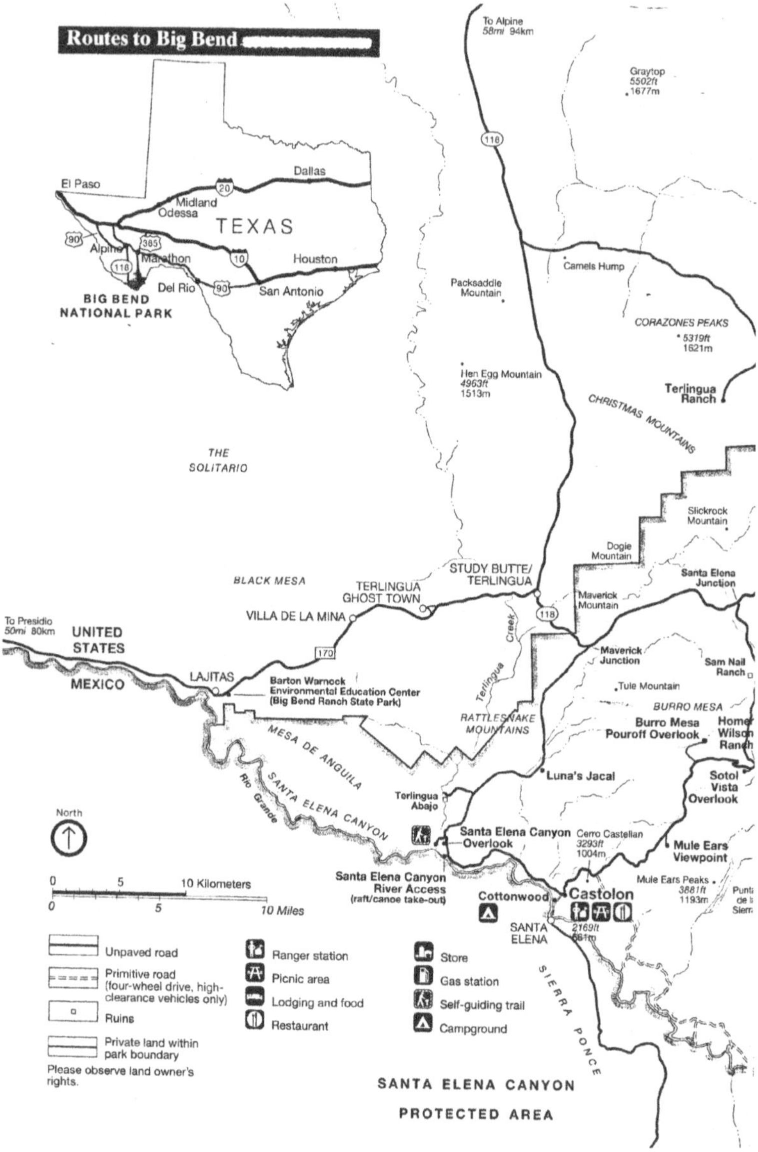
Routes to Big Bend
El Paso
Dallas
Midland
Odessa
TEXAS
Alpine
Marathon
Houston
Del Rio
San Antonio
BIG BEND NATIONAL PARK
20
90
385
10
118
To Alpine 58mi 94km
Graytop 5502ft 1677m
Camels Hump
Packsaddle Mountain
CORAZONES PEAKS 5319ft 1621m
Hen Egg Mountain 4963ft 1513m
Terlingua Ranch
CHRISTMAS MOUNTAINS
THE SOLITARIO
Slickrock Mountain
Dogie Mountain
BLACK MESA
STUDY BUTTE/ TERLINGUA
Santa Elena Junction
TERLINGUA GHOST TOWN
VILLA DE LA MINA
Maverick Mountain
To Presidio 50mi 80km
UNITED STATES
MEXICO
170
Terlingua Creek
Maverick Junction
Sam Nail Ranch
LAJITAS
Barton Warnock Environmental Education Center (Big Bend Ranch State Park)
Tule Mountain
BURRO MESA
RATTLESNAKE MOUNTAINS
Burro Mesa Pouroff Overlook
Homer Wilson Ranch
MESA DE ANGUILA
Luna's Jacal
Sotol Vista Overlook
Rio Grande
SANTA ELENA CANYON
Terlingua Abajo
North
Santa Elena Canyon Overlook
Cerro Castellan 3293ft 1004m
Mule Ears Viewpoint
0 5 10 Kilometers
0 5 10 Miles
Santa Elena Canyon River Access (raft/canoe take-out)
Mule Ears Peaks 3881ft 1193m
Cottonwood
Castolon
SANTA ELENA
2169ft
Unpaved road
Primitive road (four-wheel drive, high-clearance vehicles only)
Ruins
Private land within park boundary
Please observe land owner's rights.
Ranger station
Picnic area
Lodging and food
Restaurant
Store
Gas station
Self-guiding trail
Campground
SIERRA PONCE
SANTA ELENA CANYON PROTECTED AREA

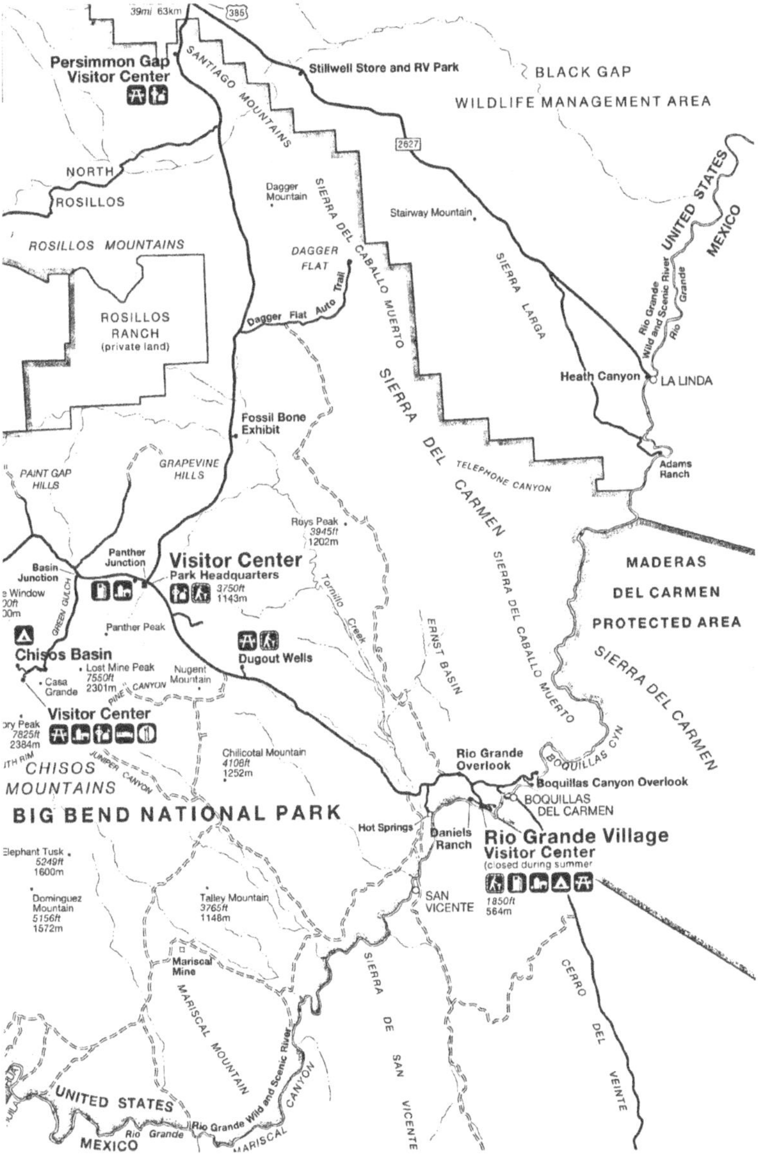
39mi 63km
385
Persimmon Gap
Visitor Center
Stillwell Store and RV Park
BLACK GAP
WILDLIFE MANAGEMENT AREA
SANTIAGO MOUNTAINS
2627
NORTH
ROSILLOS
ROSILLOS MOUNTAINS
Dagger Mountain
SIERRA DEL CABALLO MUERTO
Stairway Mountain
UNITED STATES
MEXICO
DAGGER FLAT
Dagger Flat Auto Trail
SIERRA LARGA
Rio Grande Wild and Scenic River
Rio Grande
ROSILLOS RANCH
(private land)
SIERRA DEL CARMEN
Heath Canyon
LA LINDA
Fossil Bone Exhibit
Adams Ranch
PAINT GAP HILLS
GRAPEVINE HILLS
TELEPHONE CANYON
Roys Peak
3945ft
1202m
MADERAS
DEL CARMEN
PROTECTED AREA
Panther Junction
Basin Junction
Visitor Center
Park Headquarters
3750ft
1143m
GREEN GULCH
Panther Peak
Tornillo Creek
ERNST BASIN
SIERRA DEL CABALLO MUERTO
SIERRA DEL CARMEN
Dugout Wells
Chisos Basin
Lost Mine Peak
7550ft
2301m
Casa Grande
Nugent Mountain
PINE CANYON
Visitor Center
7825ft
2384m
CHISOS
MOUNTAINS
JUNIPER CANYON
Chilicotal Mountain
4108ft
1252m
Rio Grande Overlook
BOQUILLAS CYN
Boquillas Canyon Overlook
BOQUILLAS
DEL CARMEN
BIG BEND NATIONAL PARK
Hot Springs
Daniels Ranch
Rio Grande Village
Visitor Center
(closed during summer
1850ft
564m
Elephant Tusk
5249ft
1600m
SAN VICENTE
Dominguez Mountain
5156ft
1572m
Talley Mountain
3765ft
1148m
Mariscal Mine
CERRO DEL VEINTE
SIERRA DE SAN VICENTE
MARISCAL MOUNTAIN
UNITED STATES
MEXICO
Rio Grande
Rio Grande Wild and Scenic River
MARISCAL CANYON

## Approaching Big Bend

Big Bend National Park sweeps so far south that, since there is no road through it to Mexico, the park isn't on the way to anywhere. That makes an automobile the best transportation bet. Trains and transcontinental buses approach only as close as Alpine and Marathon (see map). There is no public transportation to or through the park. You can fly into the Midland-Odessa airport to the north. Cars may be rented in Alpine and in the Midland and Odessa areas. The distances encountered are vast, so plan departures and arrivals conveniently for available facilities.

From San Antonio, Texas, it is 660 kilometers (410 miles) to the park headquarters at Panther Junction via U.S. 90 to Marathon and south on U.S. 385. Driving from the Persimmon Gap park entrance to Panther Junction will consume 46 of these kilometers (29 miles). The gap is a low mountain pass once traversed by the Comanche War Trail, a thoroughfare northward for thousands of horses stolen in Mexico. Had you crossed this pass in 1859 you might have witnessed the U.S. military experimenting with camels as beasts of burden for this dry country. The cantankerous camels bettered the standard military mule on several points. However, the Civil War ended the camel tryouts. The camel's aptness illustrates that you are traveling desert country. This calls for unique precautions, so please read the Tips for Desert Travelers on page 124.

From El Paso it is 520 kilometers (323 miles) to Panther Junction via Interstate 10 to Van Horn, U.S. 90 to Alpine, and south via Texas 118. You enter the park at the Maverick entrance. Just west of the park here on Texas Ranch Road 170 is the ghost town of Terlingua, a worthwhile side trip. In the park's western section you find the Painted Desert, eroded badlands formations showing distinct bands of colorful deposits from up to 70 million years ago.

From El Paso and points west you can take U.S. 67 south from Marfa to Presidio, approaching the park on Texas Ranch Road 170, the Camino del Rio, at the Maverick entrance. From this entrance to the headquarters at Panther Junction is about 43 kilometers (27 miles).

Water and gasoline are available in and around the park only at a few and often widely separated points. Check your water supply and gas gauge before you leave U.S. 90.

**Driving the Park.** At Panther Junction you can purchase the *Road Guide to Paved and Improved Dirt Roads of Big Bend National Park*. It describes six tours and the points of interest enroute. The Old Maverick Drive, Basin Drive, and Ross Maxwell Scenic Drive begin at junctions along the park road from Maverick to Panther Junction. The Rio Grande Village Drive begins at Panther Junction. The Persimmon Gap and Maverick drives go from their respective entrances to Panther Junction. A park map and information folder includes a large map showing major natural and historical features, roads, and facilities and services.

You can obtain information before your trip by calling or writing the Superintendent, Big Bend National Park, Texas 79834, (915) 477-2251. Or find information on the Internet at www.nps.gov/bibe.

**Primitive Roads.** After you have toured the main points on the major park roads, you may want to see more by

vehicle. You can do so on the primitive roads, which introduce further varieties of scenery and interesting plants, animals, and historic features. Plan your trip in advance, don't just turn off a main road on the spur of the moment. Check in at the park headquarters for current information about road conditions. On these primitive roads you are on your own, so play it safe. The primitive roads are patrolled only infrequently. Some are suited only for 4-wheel drive. A *Road Guide to Backcountry Dirt Roads of Big Bend National Park* may be purchased at any visitor center.

**Persimmon Gap Drive.** This drive offers short side trips: a motor nature trail up to Dagger Flat, or to the Fossil Bone Exhibit located off the highway north of the Tornillo Creek bridge. A trail guide for the Dagger Flat road identifies Chihuahuan Desert species, including the giant-dagger yuccas, found in the United States only in Big Bend country. The Fossil Bone Exhibit shows an extinct mammal, *Coryphodon*, whose remains were found in sandstone deposits about 50 million years old. Tornillo Creek is one of the park's largest drainages. The Chisos Mountains, seen as you approach Panther Junction, are the park's highest. Panther Junction is such a focal point that you may overhear park employees call it PJ.

**Maverick Drive.** Terlingua and Study Butte, now sparsely populated, were once prosperous cinnabar (mercury) mining communities. The large, rounded Maverick Mountain north of the road near the Maverick entrance is the eroded exposure of an intruded mass of molten rock pushed up through softer, older rock beds. You will also see the Painted Desert and many plants of the desert shrub community. Outside the park to the north the Christmas Mountains are prominent. To the east the Chisos define the skyline. Along the Maverick drive you come to the Santa Elena Junction, where the Ross Maxwell Scenic Drive begins (see below). Further on you will see dry washes that can carry flash floods and gravel slopes formed by the erosion of the mountains. Then you come to the Basin Junction, where the Basin drive begins (see below). Near this area you may also see mule deer, the peccary (or javelina), coyote, or other desert animals. The next stop, except for taking pictures, is Panther Junction.

**Rio Grande Village Drive.** From Panther Junction you can head southeast toward the Rio Grande's Boquillas Canyon. Along the way are the Dugout Wells picnic area and self-guiding nature trail, Hot Springs, and Rio Grande Village (see Facilities and Services). Boquillas Canyon is one of the Rio Grande's three grandest canyons here in the park. It was cut through the Sierra del Carmen (*sierra* is Spanish for mountains) and is the longest of Big Bend's famous gorges. Across the river is the Mexican village of Boquillas.

**Ross Maxwell Scenic Drive.** Spectacular historic and geologic features are found along the Ross Maxwell Scenic Drive. You observe wall-like dikes, massive gravel deposits, an ancient buried valley, and a narrow canyon cut through volcanic tuff. Across the river near Castolon is the Mexican village of Santa Elena. On the U.S. side are adobe and stone ruins of dwellings for farms on the river flats.

Near the end of this drive a viewpoint gives an excellent view of Santa Elena Canyon. Summer sunlight only strikes the canyon mouth for several hours after sunrise. To take pictures, make this trip in early morning. The canyon is usually hot in midday during summer.

**Basin Drive.** From Basin Junction southward the Basin drive climbs out of desert lowlands into the woodlands of the Chisos Mountains and their Basin, the park's "island" of green. The grade of the approach road is deceptive: watch that your vehicle does not overheat. The Big Bend agave plant reaches heights of 4 meters (15 feet). Even if you miss its bright yellow blossoms in summer, the old stalk remains standing for a year or two. As you drive up Green Gulch the vegetation changes from desert shrub to woodland species. The road's highest point is Panther Pass, more than 1.6 kilometers (1 mile) in elevation. At the parking lot here the Lost Mine Trail begins its ascent of Lost Mine Ridge, where legends of a rich Spanish mine have touched off many a vain (no pun intended) search. From the top you get superb views into Mexico. As you leave the parking lot the road begins to descend into the Chisos Basin via a series of sharp curves. (Read about the Chisos Basin under Facilities and Services.)

**Driving Safety.** Be aware that there are desert-related driving hazards not mentioned above. Please see Tips for Desert Travelers.

## Facilities and Services

Overnight lodging is available inside the park at the Chisos Mountains Lodge in the Basin; at campgrounds at the Basin, Castolon, and Rio Grande Village; and at a small trailer park at Rio Grande Village.

Chisos Mountains Lodge accommodates more than 200 guests in stone-and-adobe cottages with terraces and private baths and in modern motel-type units. Reservations should be made by writing to National Park Concessions, Inc., Big Bend National Park, Texas 79834. The lodge restaurant serves food from 7 a.m. to 8 p.m. daily.

**Campgrounds.** Campgrounds at the Basin, Castolon, and Rio Grande Village provide water and comfort stations. Electricity and fuel are not available, except that the concessioner sells charcoal. Ground fires are not permitted. Only at Rio Grande Village are there pay showers and laundry facilities for campers. Organized groups of 10 or more may make advance campground reservations, but families, other small parties, and individuals may not.

A number of backcountry roadside campsites are located along the park's backcountry dirt roads. A free permit, available at any visitor center, is required. If you are interested in a primitive site, ask at a visitor center.

**Trailer Parks.** Rio Grande Village trailer park has hookups; use of all hookups is required. Consult a ranger before taking trailers into the Basin campground. The road to the Basin is discouraged for trailers longer than 6 meters (20 feet) or RVs (7 meters/24 feet) due to sharp curves and steep grades.

**Stores and Service Stations.** Groceries, cold drinks, camping supplies, and

*A rainbow cactus in bloom (top) is about to be immortalized by a young photographer. The Panther Path (bottom), a self-guiding nature trail outside the visitor center, introduces you to cactuses and other desert plants of the park.*

film can be bought at the Basin, Rio Grande Village, Castolon, and Panther Junction. There is a gift shop in the Chisos Mountains Lodge. Minor auto repair service and gasoline may be obtained at Panther Junction. Gasoline is sold in the park at Rio Grande Village and Panther Junction only, so check your gauge before you leave U.S. 90, and periodically in the park.

**Interpretive Programs.** Illustrated evening talks are presented in summer at dusk at the amphitheater in the Basin, and in other seasons at the Basin and Rio Grande Village. Schedules for all programs are listed on park bulletin boards.

The interpretive programs at Big Bend are designed to help you understand and enjoy the natural and historic features. The park road system features wayside exhibits and markers about local attractions. The Dagger Flat Interpretive Auto Trail leaves the park road south of Persimmon Gap. Self-guiding trails are located at Panther Junction, Dugout Wells, Santa Elena Canyon, the Basin, and Rio Grande Village. A descriptive booklet is available at the head of the Lost Mine Trail in the Chisos Mountains, to help you identify plants, animals, and panoramic views.

Park rangers provide many other services to interpret the park's geology, plants, animals, and history. You can supplement these personal services by purchasing maps and publications at any of the park's four visitor centers.

## Walking Trips and Hiking

Besides floating the river, the surest way to experience the "real time" of the park is to walk out into it. Neither desert nor mountains will reveal themselves to a motor vehicle. Opportunities for walks and hikes abound, from short strolls on well-developed nature trails to multi-day, cross-park treks via its network of trails. Walks and hikes are described in *Hiker's Guide to the Developed Trails and Primitive Routes, Big Bend National Park.* Hiking guides and topographic maps sold at park visitor centers, by mail, or on the Internet, give particulars. (See Armchair Explorations.) General information follows here.

If you take a trail, please stay on it. Trails are routed for safety and constructed to avoid washouts and erosion. Shortcutting increases erosion and rockslides. Sturdy street shoes or sneakers will suffice on short developed trails. Wear hiking boots for longer hikes to protect against sharp rocks and the spines of desert plants. Carry tweezers in your first aid kit, for pulling spines and thorns. Take plenty of food and water to carry you through your return or your connection with new supplies. In hot weather one person needs 4 liters (1 gallon) of water per day; in winter slightly less. In winter rattlesnakes are rare. In summer they are common, but are mostly abroad at night. (See Tips for Desert Travelers.)

Smoking is prohibited on trails because fire poses a real threat to plants and animals in this dry country. Building ground fires is prohibited too. You will need charcoal or a camping stove in the campgrounds. In the backcountry you will need a backpacking stove and sufficient fuel for cooking and for heating water.

Water is a precious resource anywhere, but here in Big Bend it is also scarce. Do not pollute streams, springs, or tinajas by washing in them or close to them. Also be careful where you make your toilet in the backcountry.

Short walks on developed trails are available throughout the park. Short walks of 3.2 kilometers (2 miles) or less include: Window View Trail (Chisos Basin); Rio Grande Village Nature Trail; Burro Mesa Pouroff Trail (Ross Maxwell Scenic Drive); Boquillas Canyon Trail; Chisos Basin Loop Trail; Santa Elena Canyon Trail; and Hot Springs Trail (off the road to Rio Grande Village).

Beyond the short walks, hiking and backcountry options are endless. If you go off trail, wear proper gear, carry adequate supplies, use a topographic map, and know your route. Get advice on routes and gear from any park ranger or at a visitor center. It is important that you inform someone, preferably a friend or relative, of your intended route. A backcountry permit is required whenever you plan to camp overnight anywhere but in developed campgrounds. Backcountry permits are issued free at park visitor centers (see map).

Take it easy on the trails and enjoy yourself, especially until you are acclimated. In the mountains, the elevation adds to the exertion. In the lowlands the heat is an important factor.

## Birding

To learn about the raven, writes Barry Lopez in *Desert Notes*, "bury yourself in the desert so that you have a commanding view of the high basalt cliffs where he lives. Let only your eyes protrude. Do not blink—the movement will alert the raven to your continued presence . . . . there will be at least one bird who will find you. He will see your eyes staring up out of the desert floor. The raven is cautious, but he is thorough. He will sense your peaceful intentions. Let him have the first word. Be careful: he will tell you he knows nothing."

The raven is one of nearly 450 species of birds that have been seen in Big Bend National Park. Why so many? The park lies right smack on the flyway for birds winging north out of Mexico and, indeed, almost three-fourths of the species recorded in the park fly right on through. But for serious birders—even those less determined than the raven seeker Lopez describes—this is paradise. During the spring migration northbound birds confront the northwest-southeast trending Sierra del Carmen. Instead of flying over them, most birds keep to the west and are funneled right into the park. For the return trip in fall, however, the Sierra del Carmen have the opposite effect. Many southbound birds peel off east at Persimmon Gap. The fall migration is smaller and brings more lowland than mountain species.

The springtime bursts of birdsong are inspired by the drive to claim territory, as some 100 species nest within the park. Many of these breeding birds will abandon the lowlands as soon as their young are big enough, moving to the mountains to beat the heat. Rio Grande Village, for example, is at its low ebb for birdlife in July, and in the Chisos Basin you will find yourself identifying lowland birds left and right! But all in all, Rio Grande Village, with its ponds, cottonwood groves, rich riverbottoms, heavy brushlands, and neighboring desert, is the best year-round birding site in the park. In the springtime it offers an unbelievable display of species.

The Chisos Basin is the second best overall birding site, providing a long parade of mountain birds throughout the year, as well as lowland birds. The Basin is one of two spots for the annual Christmas Bird Count, along with Rio Grande Village. Over a five-year span 147 species have been counted at these locations at Christmas time.

The Colima warbler is probably the park's most famous bird, because in all the United States it nests only here in the Chisos Mountains. It arrives in April from southwest Mexico and leaves in mid-September. The Big Bend area is also a breeding ground of the peregrine falcon, a species that suffered much from pesticides. Peregrines nest in the Chisos Mountains and in the canyons along the Rio Grande. Although peregrine falcon populations have increased nationwide in recent years, nesting success in the southwestern United States, including Big Bend, tends to be poor. These predatory birds prefer to nest in a scrape on a high cliff ledge. This falcon, about the size of a crow, is slaty-backed and barred below, with a pair of black "mustaches" on the face. In its magnificent dive, the peregrine is one of the fastest moving animals on Earth.

For advice on good birding spots at any time of the year, or on where to find specific birds, ask a ranger at Panther Junction. (Also see Armchair Explorations.)

## Birds of Big Bend

*Summer tanager*

*Blackthroated sparrow*

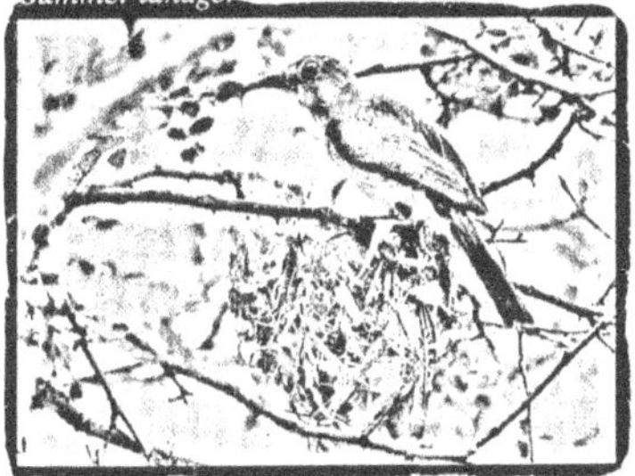

*Gray vireo*

*Painted bunting*

*Great horned owl*

*Belted kingfisher*

*Yellowbreasted cha*

*Roadrunne*

## Floating the River

*Boulders dwarf the raft of Park Service employees (top) landing in Santa Elena Canyon. A kayaker (bottom) drifts into reflections.*

If you stumbled onto the Rio Grande upstream of the park, between El Paso and Presidio, during most of the year you'd say "Oh well, forget floating!" That stretch is most often dry, sapped by irrigation projects. But the Rio Grande gets a new lease on life as the Rio Conchos, draining Mexican mountains, flows into it at Presidio. You can thank the Rio Conchos for the prospects of floating the Big Bend. Along part of the park boundary and for 127 miles downstream, the Rio Grande is designated a national wild and scenic river for 315 kilometers (196 miles). The "scenic" goes without saying. When you hit the rapids or a cross-channel current hits you, the "wild" designation rings true as well.

Below is general information about floating the river. Particulars—including descriptions of the canyons and some rapids—are contained in, among others, the river guide series published by the Big Bend Natural History Association. (See Armchair Explorations.)

You need a permit to float the Rio Grande in the park and along the Rio Grande Wild and Scenic River. The free permit is available at any visitor center. Water levels change daily, so ask about current conditions. Permits will not be issued if the river is at flood stage. Dangerous flash floods are most likely during summer and early fall. If you float the river in summer, be prepared for extreme temperatures. From late April through October, the heat along the river can be intense.

If river levels are high enough, the recommended craft is an inflatable raft which is not as prone to damage from submerged rocks as is a canoe. However, low water is often a problem. At times, the river is so low that it may be passable only by canoe or kayak. No craft or accessory gear is available for rental within the park. You must bring your own, or make arrangements with an outfitter. Any float party should consist of two or more people for obvious safety reasons, and everybody should be able to swim.

You will need two vehicles, one for put-in and one for take-out, or you can arrange a shuttle from a friend or a professional shuttle service. Gear should include: approved personal flotation device (pfd) for each person and an extra pfd in each boat, boat paddle for each person and an extra paddle in each boat, waterproof duffle, fresh water, flashlight, lash lines and 15 meter (50 foot) bow and stern lines, first-aid kit, and boat patching kit. A dump-station compatible toilet system is required for overnight trips in the park's canyons. Lash all these items to prevent their loss in the event that your craft capsizes.

Lest you be discouraged by these safety warnings and logistical considerations, suffice it to say that from the river inside Big Bend's magnificent gorges you will experience an intimate immensity rare on this Earth.

## Fishing

Most of the park's native fish are of minnow size, but the Rio Grande does attract anglers. The complete list of fish recorded in the park and its immediate surroundings includes 40 species, including bluegill and sunfish species.

Most anglers are after the blue, channel, and flathead catfish. The blue and flathead are favored food fish. The longnose gar may reach over a meter (4 feet) in length and is predatory, as its long snout and sharp teeth suggest. You do not need a state fishing license to fish in the national park. A free park fishing permit can be obtained at any visitor center. For advice on fishing spots and preferred methods, ask a park ranger.

Yellow cat up to 45 kilos (100 pounds) have been taken from the river, and 14-kilo (30-pound) cats are not uncommon. Channel and blue cats also provide fine sport and good eating. These deepwater species feed on aquatic plants, insects, and smaller fish, both living and dead. They spawn in depressions and sheltered nooks in river banks and canyon cliffs. While catfish account for most of the recreational fishing in the Rio, many other interesting species swim the brown waters: the predatory garfish, needlenosed and shaped like a torpedo; the humpbacked carp that can survive even in limited waters; smallmouth buffalo, sheepshead, and green sunfish; the bullhead that favors quiet waters and can endure higher temperatures and lower oxygen content than most other fish; and of course the minnows, as plentiful and gregarious as sparrows, with a preference for running water and rocky or sandy bottoms.

## Tips for Desert Travelers

**Best Times to Visit Big Bend.** Winter days will be nippy in Big Bend's mountains and comfortably warm in the lowlands. Snow falls in the mountains once or twice a year. Spring hits the desert lowlands in February and begins a slow ascent up the mountain heights, arriving in April. Some of the desert plants bloom throughout the year, but the most predictable displays are in springtime. Summer is problematic. If you come in the summer you will probably want to take to the mountains. Midsummer daytime temperatures in the desert and river valley generally hover above 38°C (100°F)—often well above. But up in the Basin, daytime temperatures average 29°C (85°F) and nights are cool. Autumn is usually sunny, with the air gentle and warm. The best months for running the river are November through February, both for river conditions and for avoiding summer's intense heat in the canyons.

**Climbing Hazards.** The character of the basic rock in the park is very unstable, making it unsafe for climbing. Climbing is not recommended.

**Night Driving.** Be extra alert for wildlife while you are driving at night. Many creatures, particularly deer, may be blinded by your headlights so that they make no effort to get off the road.

**Spines and Thorns.** Beware the cactus and other spine- and thorn-bearing plants, shrubs, and trees. These can inflict painful injury. To protect yourself, wear stout shoes and tough clothing if you go hiking off the developed trails. If you go out at night, walk carefully, carry a flashlight, and don't venture out too far. The National Park Service recommends that you carry tweezers for removing irritating spines, which may be too small to extract with your fingers. People have even gotten mouthfuls of delicate—but highly irritating—spines of some pricklypear cactus species while trying to eat the fruits! Delicate spines hardly noticeable to the naked eye can inflict painful injury.

**Poisonous Reptiles.** The park contains copperhead snakes and four species of rattlesnakes, although these are seldom seen in daylight. Though poisonous, they are protected in the park. Do not molest or harm them. Very few snakebites occur in the park. Most of these involve bites to the hands of people who have reached into places where they couldn't see a snake resting or hiding, and to people walking at night without flashlights. Stay on trails after dark and use a flashlight, and avoid bushes and damp areas. There are no poisonous lizards at Big Bend, but if you molest lizards they may bite you.

**Tarantulas and Scorpions.** Tarantulas, contrary to horror films, will not bite you unless you annoy them. And the park species of scorpion is not deadly.

**No Medical Services.** There are no doctors or nurses in the park, but many park rangers are trained as Emergency Medical Technicians. The closest hospital, in Alpine, is 161 kilometers (100 miles) from Panther Junction. A small clinic is located in Study Butte. Carry your own first-aid supplies (including tweezers). Notify the nearest park ranger immediately of accidents or emergencies.

**Use Water Wisely.** Carry drinking water with you whenever you are in desert country. This is particularly a must when hiking: 4 liters (1 gallon) per day per person is a good rule of thumb. And remember: Start your return trip *before* half of your water supply is used up. Floaters and boaters, do not drink the river water: Carry your own. Any spring water used for drinking should be treated first.

**Flash Floods.** High water is a threat during the flash floods that may follow summer thunderstorms. You must be alert for this because a flash flood may travel down the watershed to you even though you were not rained on. Flash floods make roadway dips potential death traps. If you are caught in high water, drive slowly to avoid stalling your engine: Do not splash through at a high speed. By all means, avoid any high water! Washouts are a hazard in a storm and afterwards, so be on the lookout for them. These same precautions about high water and low spots apply to camping too. Do not camp in washes and arroyos. They could turn into swirling rivers while you sleep, and you might not even have the warning of raindrops to wake you.

## Park Regulations

**Driving.** The maximum speed on park roads is 72 kilometers (45 miles) per hour. Slower limits are posted; please observe them. Do not pass or park on curves, and take road dips slowly. Motorized vehicles are restricted to the park roadways and are prohibited on all trails. Check with a park ranger before driving the primitive roads.

**Trail Use.** Stay on trails because shortcuts mar the area and can cause erosion and destructive rockslides. Smoking is not allowed on trails because fire is a constant hazard here. Please carry out all your refuse.

**Camping and Fires.** Camping is limited to campgrounds except for backcountry camping, which is by permit only. Building wood and ground fires is prohibited.

**Fishing Licenses.** Fishing licenses are not required. Obtain fishing permits and information at any visitor center.

**Firearms and Pets.** Use or display of firearms is prohibited. Pets must be kept on a leash at all times, and they are not permitted on trails or in public buildings.

**River Use.** A permit is required to float the Rio Grande. At high water float trips are particularly hazardous. No permits are issued for Santa Elena and Mariscal Canyons when the river is at flood stage. Swimming in the river is discouraged because of the dangers, many of which are unseen. Obtain float permits at any visitor center. For information call or write the Superintendent, Big Bend National Park, Texas 79834, (915) 477-2251. Or on the Internet: www.nps.gov/bibe.

## Armchair Explorations

In addition to the publications listed below, books, maps, guides, and other publications are offered for sale at the park or through the mail by the Big Bend Natural History Association, P.O. Box 196, Big Bend National Park, Texas 79834. Call (915) 477-2236 for a free catalog. Or shop on the Internet at www.bigbendbookstore.org.


*Deserts*, National Audubon Society Nature Guide. Alfred K. Knopf, 1997.

Gomez, Arthur R., *A Most Singular Country—A History of Occupation in the Big Bend*. Signature Books, 1990.

Langford, J. O. with Fred Gipson. *Big Bend, A Homesteader's Story*. University of Texas Press, 1973.

Larson, Peggy. *The Deserts of the Southwest*. Sierra Club Books, 1977.

Maxwell, Ross A. *The Big Bend of the Rio Grande*. Guidebook 7. Texas Bureau of Economic Geology, 1968.

Ragsdale, Kenneth B., *Big Bend Country, Land of the Unexpected*. University of Texas Press, 1998.

Sonnichsen, C. L. *The Mescalero Apaches* (Second edition). University of Oklahoma Press, 1973.

Tyler, Ronnie C. *The Big Bend: A History of the Last Texas Frontier*. Texas A&M Press, 1996.

Wauer, Roland H. *A Field Guide to the Birds of the Big Bend*. Gulf Publishing, 1996.

Wauer, Roland H. *Naturalists' Big Bend*. Texas A&M Press, 1980.

# Index

*Numbers in italics refer to photographs, illustrations, or maps.*




☆GPO: 2000—460-794/00509
Reprint 2000.

Spotted something that isn't quite right?

Let us know and we'll fix it for future reprints. And as a thank you for your diligence, we'll send you a free Sothis book of your choice.

publisher@sothis.press

This page intentionally left blank

Made in United States
Cleveland, OH
05 November 2026

47668299R00080